AAM8546
UL FMT B RT a BL m T/C DT 07/15/88 R/DT 08/01/90 CC STAT mm E/L 1
010: : |a 14007045
035/1: : |a (RLIN)MIUG83-B73555
035/2: : |a (CaOTULAS)160088910
040: : |c MiU |d MiU
050/1:0 : |a QA241 |b .C3
100:1 : |a Carmichael, Robert Daniel, |d 1879-
245:04: |a The theory of numbers, |c by Robert D. Carmichael ...
250: : |a 1st ed., 1st thousand.
260: : |a New York, |b J. Wiley & sons, inc.; [etc., etc.] |c 1914.
300/1: : |a 94 p. |c 24 cm.
490/1:1 : |a Mathematical monographs. Ed. by M. Merriman and R. S. Woodward. no. 13
650/1: 0: |a Number theory.
830/1: 0: |a Mathematical monographs |v no. 13.
998: : |c EM |s 9124

MATHEMATICAL MONOGRAPHS.

EDITED BY

MANSFIELD MERRIMAN AND ROBERT S. WOODWARD.

No. 13.

THE THEORY

OF

NUMBERS

BY

ROBERT D. CARMICHAEL,
Associate Professor of Mathematics in Indiana University

FIRST EDITION
FIRST THOUSAND

NEW YORK
JOHN WILEY & SONS, Inc.
London: CHAPMAN & HALL: Limited
1914

THE SCIENTIFIC PRESS
ROBERT DRUMMOND AND COMPANY
BROOKLYN, N. Y.

EDITORS' PREFACE.

THE volume called Higher Mathematics, the third edition of which was published in 1900, contained eleven chapters by eleven authors, each chapter being independent of the others, but all supposing the reader to have at least a mathematical training equivalent to that given in classical and engineering colleges. The publication of that volume was discontinued in 1906, and the chapters have since been issued in separate Monographs, they being generally enlarged by additional articles or appendices which either amplify the former presentation or record recent advances. This plan of publication was arranged in order to meet the demand of teachers and the convenience of classes, and it was also thought that it would prove advantageous to readers in special lines of mathematical literature.

It is the intention of the publishers and editors to add other monographs to the series from time to time, if the demand seems to warrant it. Among the topics which are under consideration are those of elliptic functions, the theory of quantics, the group theory, the calculus of variations, and non-Euclidean geometry; possibly also monographs on branches of astronomy, mechanics, and mathematical physics may be included. It is the hope of the editors that this Series of Monographs may tend to promote mathematical study and research over a wider field than that which the former volume has occupied.

PREFACE

The purpose of this little book is to give the reader a convenient introduction to the theory of numbers, one of the most extensive and most elegant disciplines in the whole body of mathematics. The arrangement of the material is as follows: The first five chapters are devoted to the development of those elements which are essential to any study of the subject. The sixth and last chapter is intended to give the reader some indication of the direction of further study with a brief account of the nature of the material in each of the topics suggested. The treatment throughout is made as brief as is possible consistent with clearness and is confined entirely to fundamental matters. This is done because it is believed that in this way the book may best be made to serve its purpose as an introduction to the theory of numbers.

Numerous problems are supplied throughout the text. These have been selected with great care so as to serve as excellent exercises for the student's introductory training in the methods of number theory and to afford at the same time a further collection of useful results. The exercises marked with a star are more difficult than the others; they will doubtless appeal to the best students.

Finally, I should add that this book is made up from the material used by me in lectures in Indiana University during the past two years; and the selection of matter, especially of exercises, has been based on the experience gained in this way.

R. D. Carmichael.

CONTENTS

CHAPTER I. ELEMENTARY PROPERTIES OF INTEGERS

CHAPTER II. ON THE INDICATOR OF AN INTEGER

CHAPTER III. ELEMENTARY PROPERTIES OF CONGRUENCES

CHAPTER IV. THE THEOREMS OF FERMAT AND WILSON

CHAPTER V. PRIMITIVE ROOTS MODULO m

CHAPTER VI. OTHER TOPICS

THE THEORY OF NUMBERS

CHAPTER I

ELEMENTARY PROPERTIES OF INTEGERS

§ 1. Fundamental Notions and Laws

In the present chapter we are concerned primarily with certain elementary properties of the positive integers 1, 2, 3, 4, ... It will sometimes be convenient, when no confusion can arise, to employ the word *integer* or the word *number* in the sense of positive integer.

We shall suppose that the integers are already defined, either by the process of counting or otherwise. We assume further that the meaning of the terms *greater*, *less*, *equal*, *sum*, *difference*, *product* is known.

From the ideas and definitions thus assumed to be known follow immediately the theorems:

I. *The sum of any two integers is an integer.*
II. *The difference of any two integers is an integer.*
III. *The product of any two integers is an integer.*

Other fundamental theorems, which we take without proof, are embodied in the following formulas:

IV. $$a+b=b+a.$$
V. $$a\times b=b\times a.$$
VI. $$(a+b)+c=a+(b+c).$$
VII. $$(a\times b)\times c=a\times(b\times c).$$
VIII. $$a\times(b+c)=a\times b+a\times c.$$

Here a, b, c denote any positive integers.

These formulas are equivalent in order to the following five theorems: addition is commutative; multiplication is commutative; addition is associative; multiplication is associative; multiplication is distributive with respect to addition.

EXERCISES

1. Prove the following relations:

$$1+2+3+\ldots+n=\frac{n(n+1)}{2},$$

$$1+3+5+\ldots+(2n-1)=n^2,$$

$$1^3+2^3+3^3+\ldots+n^3=\left(\frac{n(n+1)}{2}\right)^2=(1+2+\ldots+n)^2.$$

2. Find the sum of each of the following series:

$$1^2+2^2+3^2+\ldots+n^2,$$

$$1^2+3^2+5^2+\ldots+(2n-1)^2,$$

$$1^3+3^3+5^3+\ldots+(2n-1)^3.$$

3. Discover and establish the law suggested by the equations $1^2=0+1$, $2^2=1+3$, $3^2=3+6$, $4^2=6+10$, . . . ; by the equations $1=1^3$, $3+5=2^3$, $7+9+11=3^3$, $13+15+17+19=4^3$, ...

§ 2. Definition of Divisibility. The Unit

Definitions. An integer a is said to be divisible by an integer b if there exists an integer c such that $a=bc$. It is clear from this definition that a is also divisible by c. The integers b and c are said to be divisors or factors of a; and a is said to be a multiple of b or of c. The process of finding two integers b and c such that bc is equal to a given integer a is called the process of resolving a into factors or of factoring a; and a is said to be resolved into factors or to be factored.

We have the following fundamental theorems:

I. *If b is a divisor of a and c is a divisor of b, then c is a divisor of a.*

Since b is a divisor of a there exists an integer β such that $a=b\beta$. Since c is a divisor of b there exists an integer γ such that $b=c\gamma$. Substituting this value of b in the equation $a=b\beta$ we have $a=c\gamma\beta$. But from theorem III of § 1 it follows that $\gamma\beta$ is an integer; hence, c is a divisor of a, as was to be proved.

II. *If c is a divisor of both a and b, then c is a divisor of the sum of a and b.*

From the hypothesis of the theorem it follows that integers α and β exist such that

$$a=c\alpha, \quad b=c\beta.$$

Adding, we have

$$a+b=c\alpha+c\beta=c(\alpha+\beta)=c\delta,$$

where δ is an integer. Hence, c is a divisor of $a+b$.

III. *If c is a divisor of both a and b, then c is a divisor of the difference of a and b.*

The proof is analogous to that of the preceding theorem.

DEFINITIONS. If a and b are both divisible by c, then c is said to be a common divisor or a common factor of a and b. Every two integers have the common factor 1. The greatest integer which divides both a and b is called the greatest common divisor of a and b. More generally, we define in a similar way a common divisor and the greatest common divisor of n integers $a_1, a_2, \ldots, a_n$.

DEFINITIONS. If an integer a is a multiple of each of two or more integers it is called a common multiple of these integers. The product of any set of integers is a common multiple of the set. The least integer which is a multiple of each of two or more integers is called their least common multiple.

It is evident that the integer 1 is a divisor of every integer and that it is the only integer which has this property. It is called the unit.

DEFINITION. Two or more integers which have no common factor except 1 are said to be prime to each other or to be relatively prime.

DEFINITION. If a set of integers is such that no two of them have a common divisor besides 1 they are said to be prime each to each.

EXERCISES

1. Prove that n^3-n is divisible by 6 for every positive integer n.

2. If the product of four consecutive integers is increased by 1 the result is a square number.

3. Show that $2^{4n+2}+1$ has a factor different from itself and 1 when n is a positive integer.

§ 3. PRIME NUMBERS. THE SIEVE OF ERATOSTHENES

DEFINITION. If an integer p is different from 1 and has no divisor except itself and 1 it is said to be a prime number or to be a prime.

DEFINITION. An integer which has at least one divisor other than itself and 1 is said to be a composite number or to be composite.

All integers are thus divided into three classes:

1) The unit;

2) Prime numbers;

3) Composite numbers.

We have seen that the first class contains only a single number. The third class evidently contains an infinitude of numbers; for, it contains all the numbers 2^2, 2^3, 2^4, ... In the next section we shall show that the second class also contains an infinitude of numbers. We shall now show that every number of the third class contains one of the second class as a factor, by proving the following theorem:

I. *Every integer greater than 1 has a prime factor.*

Let m be any integer which is greater than 1. We have to show that it has a prime factor. If m is prime there is the prime factor m itself. If m is not prime we have

$$m = m_1 m_2,$$

where m_1 and m_2 are positive integers both of which are less than m. If either m_1 or m_2 is prime we have thus obtained a prime factor of m. If neither of these numbers is prime, then write

$$m_1 = m'_1 m'_2, \quad m'_1 > 1, \quad m'_2 > 1.$$

Both m'_1 and m'_2 are factors of m and each of them is less than m_1. Either we have now found in m'_1 or m'_2 a prime factor of m or the process can be continued by separating one of these numbers into factors. Since for any given m there is evidently only a finite number of such steps possible, it is clear that we must finally arrive at a prime factor of m. From this conclusion the theorem follows immediately.

Eratosthenes has given a useful means of finding the prime numbers which are less than any given integer m. It may be described as follows:

Every prime except 2 is odd. Hence if we write down every odd number from 3 up to m we shall have in the list every prime less than m except 2. Now 3 is a prime. Leave it in the list; but beginning to count from 3 strike out every third number in the list. Thus every number divisible by 3, except 3 itself, is cancelled. Then begin from 5 and cancel every fifth number. Then begin from the next uncancelled number, namely 7, and strike out every seventh number. Then begin from the next uncancelled number, namely 11, and strike out every eleventh number. Proceed in this way up to m. The uncancelled numbers remaining will be the odd primes not greater than m.

It is obvious that this process of cancellation need not be carried altogether so far as indicated; for if p is a prime greater than $\sqrt{m}$, the cancellation of any p^{th} number from p will be merely a repetition of cancellations effected by means of another factor smaller than p, as one may see by use of the following theorem.

II. *An integer m is prime if it has no prime factor equal to or less than I, where I is the greatest integer whose square is equal to or less than m.*

Since m has no prime factor less than I, it follows from theorem I that it has no factor but unity less than I. Hence, if m is not prime it must be the product of two numbers each greater than I; and hence it must be equal to or greater than $(I+1)^2$. This contradicts the hypothesis on I; and hence we conclude that m is prime.

EXERCISE

By means of the method of Eratosthenes determine the primes less than 200.

§ 4. The Number of Primes is Infinite

I. *The number of primes is infinite.*

We shall prove this theorem by supposing that the number of primes is not infinite and showing that this leads to a contradiction. If the number of primes is not infinite there is a greatest prime number, which we shall denote by p. Then form the number

$$N = 1 \cdot 2 \cdot 3 \cdot \ldots \cdot p + 1.$$

Now by theorem I of § 3 N has a prime divisor q. But every non-unit divisor of N is obviously greater than p. Hence q is greater than p, in contradiction to the conclusion that p is the greatest prime. Thus the proof of the theorem is complete.

In a similar way we may prove the following theorem:

II. *Among the integers of the arithmetic progression* 5, 11, 17, 23, . . . , *there is an infinite number of primes.*

If the number of primes in this sequence is not infinite there is a greatest prime number in the sequence; supposing that this greatest prime number exists we shall denote it by p. Then the number N,

$$N = 1 \cdot 2 \cdot 3 \cdot \ldots \cdot p - 1,$$

is not divisible by any number less than or equal to p. This number N, which is of the form $6n-1$, has a prime factor. If this factor is of the form $6k-1$ we have already reached a contradiction, and our theorem is proved. If the prime is of the form $6k_1+1$ the complementary factor is of the form $6k_2-1$. Every prime factor of $6k_2-1$ is greater than p. Hence we may treat $6k_2-1$ as we did $6n-1$, and with a like result. Hence we must ultimately reach a prime factor of the form $6k_3-1$; for, otherwise, we should have $6n-1$ expressed as a product of prime factors all of the form $6t+1$—a result which is clearly impossible. Hence we must in any case reach a contradiction of the hypothesis. Thus the theorem is proved.

The preceding results are special cases of the following more general theorem:

III. *Among the integers of the arithmetic progression a, $a+d$, $a+2d$, $a+3d$, . . . , there is an infinite number of primes, provided that a and d are relatively prime.*

For the special case given in theorem II we have an elementary proof; but for the general theorem the proof is difficult. We shall not give it here.

EXERCISES

1. Prove that there is an infinite number of primes of the form $4n-1$.

2. Show that an odd prime number can be represented as the difference of two squares in one and in only one way.

3. The expression $m^p - n^p$, in which m and n are integers and p is a prime, is either prime to p or is divisible by p^2.

4. Prove that any prime number except 2 and 3 is of one of the forms $6n+1$, $6n-1$.

§ 5. The Fundamental Theorem of Euclid

If a and b are any two positive integers there exist integers q and r, $q \geqq 0$, $0 \leqq r < b$, such that

$$a = qb + r.$$

If a is a multiple of b the theorem is at once verified, r being in this case 0. If a is not a multiple of b it must lie between two consecutive multiples of b; that is, there exists a q such that

$$qb < a < (q+1)b.$$

Hence there is an integer r, $0 < r < b$, such that $a = qb + r$. In case b is greater than a it is evident that $q = 0$ and $r = a$. Thus the proof of the theorem is complete.

§ 6. Divisibility by a Prime Number

I. *If p is a prime number and m is any integer, then m either is divisible by p or is prime to p.*

This theorem follows at once from the fact that the only divisors of p are 1 and p.

II. *The product of two integers each less than a given prime number p is not divisible by p.*

Let a be a number which is less than p and suppose that b is a number less than p such that ab is divisible by p, and let b be the least number for which ab is so divisible. Evidently there exists an integer m such that

$$mb < p < (m+1)b.$$

Then $p - mb < b$. Since ab is divisible by p it is clear that mab is divisible by p; so is ap also; and hence their difference $ap - mab$, $= a(p - mb)$, is divisible by p. That is, the product of a by an integer less than b is divisible by p, contrary to the assumption that b is the least integer such that ab is divisible by p. The assumption that the theorem is not true has thus led to a contradiction; and thus the theorem is proved.

III. *If neither of two integers is divisible by a given prime number p their product is not divisible by p.*

Let a and b be two integers neither of which is divisible by the prime p. According to the fundamental theorem of Euclid there exist integers m, n, α, β such that

$$a = mp + \alpha, \quad 0 < \alpha < p,$$

$$b = np + \beta, \quad 0 < \beta < p.$$

Then $$ab = (mp+\alpha)(np+\beta) = (mnp + \alpha + \beta)p + \alpha\beta.$$

If now we suppose ab to be divisible by p we have $\alpha\beta$ divisible by p. This contradicts II, since α and β are less than p. Hence ab is not divisible by p.

By an application of this theorem to the continued product of several factors, the following result is readily obtained:

IV. *If no one of several integers is divisible by a given prime p their product is not divisible by p.*

§ 7. The Unique Factorization Theorem

I. *Every integer greater than unity can be represented in one and in only one way as a product of prime numbers.*

In the first place we shall show that it is always possible to resolve a given integer m greater than unity into prime factors by a finite number of operations. In the proof of theorem I, § 3, we showed how to find a prime factor p_1 of m by a finite number of operations. Let us write

$$m = p_1 m_1.$$

If m_1 is not unity we may now find a prime factor p_2 of m_1. Then we may write

$$m = p_1 m_1 = p_1 p_2 m_2.$$

If m_2 is not unity we may apply to it the same process as that applied to m_1 and thus obtain a third prime factor of m. Since $m_1 > m_2 > m_3 > \ldots$ it is clear that after a finite number of operations we shall arrive at a decomposition of m into prime factors. Thus we shall have

$$m = p_1 p_2 \ldots p_r$$

where $p_1, p_2, \ldots, p_r$ are prime numbers. We have thus proved the first part of our theorem, which says that the decomposition of an integer (greater than unity) into prime factors is always possible.

Let us now suppose that we have also a decomposition of m into prime factors as follows:

$$m = q_1 q_2 \ldots q_s.$$

Then we have

$$p_1 p_2 \ldots p_r = q_1 q_2 \ldots q_s.$$

Now p_1 divides the first member of this equation. Hence it also divides the second member of the equation. But p_1 is prime; and therefore by theorem IV of the preceding section we see that p_1 divides some one of the factors q; we suppose that p_1 is a factor of q_1. It must then be equal to q_1. Hence we have

$$p_2 p_3 \ldots p_r = q_2 q_3 \ldots q_s.$$

By the same argument we prove that p_2 is equal to some q, say q_2. Then we have

$$p_3 p_4 \ldots p_r = q_3 q_4 \ldots q_s.$$

Evidently the process may be continued until one side of the equation is reduced to 1. The other side must also be reduced to 1 at the same time. Hence it follows that the two decompositions of m are in fact identical.

This completes the proof of the theorem.

The result which we have thus demonstrated is easily the most important theorem in the theory of integers. It can also be stated in a different form more convenient for some purposes:

II. *Every non-unit positive integer m can be represented in one and in only one way in the form*

$$m = p_1^{\alpha_1} p_2^{\alpha_2} \ldots p_n^{\alpha_n}$$

where $p_1, p_2, \ldots, p_n$ are different primes and $\alpha_1, \alpha_2, \ldots, \alpha_n$ are positive integers.

This comes immediately from the preceding representation of m in the form $m = p_1 p_2 \ldots p_r$ by combining into a power of p_i all the primes which are equal to p_i.

COROLLARY 1. *If a and b are relatively prime integers and c is divisible by both a and b, then c is divisible by ab.*

COROLLARY 2. *If a and b are each prime to c then ab is prime to c.*

COROLLARY 3. *If a is prime to c and ab is divisible by c, then b is divisible by c.*

§ 8. THE DIVISORS OF AN INTEGER

The following theorem is an immediate corollary of the results in the preceding section:

I. *All the divisors of m,*

$$m = p_1^{\alpha_1} p_2^{\alpha_2} \ldots p_n^{\alpha_n},$$

are of the form

$$p_1^{\beta_1} p_2^{\beta_2} \ldots p_n^{\beta_n}, \quad 0 \leq \beta_i \leq \alpha_i;$$

and every such number is a divisor of m.

From this it is clear that every divisor of m is included once and only once among the terms of the product

$$(1+p_1+p_1^2+\ldots+p_1^{\alpha_1})(1+p_2+p_2^2+\ldots+p_2^{\alpha_2})\ldots(1+p_n+p_n^2+\ldots+p_n^{\alpha_n}),$$

when this product is expanded by multiplication. It is obvious that the number of terms in the expansion is $(\alpha_1+1)(\alpha_2+1)\ldots(\alpha_n+1)$. Hence we have the theorem:

II. *The number of divisors of m is* $(\alpha_1+1)(\alpha_2+1)\ldots(\alpha_n+1)$.

Again we have

$$\prod_i(1+p_i+p_i^2+\ldots+p_i^{\alpha_i})=\prod_i\frac{p_i^{\alpha_i+1}-1}{p_i-1}.$$

Hence,

III. *The sum of the divisors of m is*

$$\frac{p_1^{\alpha_1+1}-1}{p_1-1}\cdot\frac{p_2^{\alpha_2+1}-1}{p_2-1}\cdot\ldots\cdot\frac{p_n^{\alpha_n+1}-1}{p_n-1}.$$

In a similar manner we may prove the following theorem:

IV. *The sum of the hth powers of the divisors of m is*

$$\frac{p_1^{h(\alpha_1+1)}-1}{p_1^h-1}\cdot\ldots\cdot\frac{p_n^{h(\alpha_n+1)}-1}{p_n^h-1}.$$

EXERCISES

1. Find numbers x such that the sum of the divisors of x is a perfect square.

2. Show that the sum of the divisors of each of the following integers is twice the integer itself: 6, 28, 496, 8128, 33550336. Find other integers x such that the sum of the divisors of x is a multiple of x.

3. Prove that the sum of two odd squares cannot be a square.

4. Prove that the cube of any integer is the difference of the squares of two integers.

5. In order that a number shall be the sum of consecutive integers, it is necessary and sufficient that it shall not be a power of 2.

6. Show that there exist no integers x and y (zero excluded) such that $y^2=2x^2$. Hence, show that there does not exist a rational fraction whose square is 2.

7. The number $m=p_1^{\alpha_1}p_2^{\alpha_2}\ldots p_n^{\alpha_n}$, where the p's are different primes and the α's are positive integers, may be separated into relatively prime factors in 2^{n-1} different ways.

8. The product of the divisors of m is $\sqrt{m^v}$ where v is the number of divisors of m.

§ 9. The Greatest Common Factor of Two or More Integers

Let m and n be two positive integers such that m is greater than n. Then, according to the fundamental theorem of Euclid, we can form the set of equations

$$
\begin{array}{ll}
m = qn + n_1, & 0 < n_1 < n, \\
n = q_1 n_1 + n_2, & 0 < n_2 < n_1, \\
n_1 = q_2 n_2 + n_3, & 0 < n_3 < n_2, \\
\cdots & \cdots \\
n_{k-2} = q_{k-1} n_{k-1} + n_k, & 0 < n_k < n_{k-1}, \\
n_{k-1} = q_k n_k. &
\end{array}
$$

If m is a multiple of n we write $n = n_0$, $k = 0$, in the above equations.

Definition. The process of reckoning involved in determining the above set of equations is called the Euclidian Algorithm.

I. *The number n_k to which the Euclidian algorithm leads is the greatest common divisor of m and n.*

In order to prove this theorem we have to show two things:

1) That n_k is a divisor of both m and n;

2) That the greatest common divisor d of m and n is a divisor of n_k.

To prove the first statement we examine the above set of equations, working from the last to the first. From the last equation we see that n_k is a divisor of n_{k-1}. Using this result we see that the second member of next to the last equation is divisible by n_k. Hence its first member n_{k-2} must be divisible by n_k. Proceeding in this way step by step we show that n_2 and n_1, and finally that n and m, are divisible by n_k.

For the second part of the proof we employ the same set of equations and work from the first one to the last one. Let d be any common divisor of m and n. From the first equation we see that d is a divisor of n_1. Then from the second equation it follows that d is a divisor of n_2. Proceeding in this way we show finally that d is a divisor of n_k. Hence any common divisor, and in particular the greatest common divisor, of m and n is a factor of n_k.

This completes the proof of the theorem.

COROLLARY. *Every common divisor of m and n is a factor of their greatest common divisor.*

II. *Any number n_i in the above set of equations is the difference of multiples of m and n.*

From the first equation we have

$$n_1 = m - qn$$

so that the theorem is true for $i = 1$. We shall suppose that the theorem is true for every subscript up to $i-1$ and prove it true for the subscript i. Thus by hypothesis we have *

$$n_{i-2} = \pm(\alpha_{i-2}m - \beta_{i-2}n),$$
$$n_{i-1} = \mp(\alpha_{i-1}m - \beta_{i-1}n).$$

Substituting in the equation

$$n_i = -q_{i-1}n_{i-1} + n_{i-2}$$

we have a result of the form

$$n_i = \pm(\alpha_i m - \beta_i n).$$

From this we conclude at once to the truth of the theorem.

Since n_k is the greatest common divisor of m and n, we have as a corollary the following important theorem:

III. *If d is the greatest common divisor of the positive integers m and n, then there exist positive integers α and β such that*

$$\alpha m - \beta n = \pm d.$$

If we consider the particular case in which m and n are relatively prime, so that $d = 1$, we see that there exist positive integers α and β such that $\alpha m - \beta n = \pm 1$. Obviously, if m and n have a common divisor d, greater than 1, there do not exist

* If $i = 2$ we must replace n_{i-2} by n.

integers α and β satisfying this relation; for, if so, d would be a divisor of the first member of the equation and not of the second. Thus we have the following theorem:

IV. *A necessary and sufficient condition that m and n are relatively prime is that there exist integers α and β such that* $\alpha m - \beta n = \pm 1$.

The theory of the greatest common divisor of three or more numbers is based directly on that of the greatest common divisor of two numbers; consequently it does not require to be developed in detail.

EXERCISES

1. If d is the greatest common divisor of m and n, then m/d and n/d are relatively prime.

2. If d is the greatest common divisor of m and n and k is prime to n, then d is the greatest common divisor of km and n.

3. The number of multiplies of b in the sequence $a, 2a, 3a, \ldots, ba$ is equal to the greatest common divisor of a and b.

4. If the sum or the difference of two irreducible fractions is an integer, the denominators of the fractions are equal.

5. The algebraic sum of any number of irreducible fractions, whose denominators are prime each to each, cannot be an integer.

6*. The number of divisions to be effected in finding the greatest common divisor of two numbers by the Euclidian algorithm does not exceed five times the number of digits in the smaller number (when this number is written in the usual scale of 10).

§ 10. The Least Common Multiple of Two or More Integers

I. *The common multiples of two or more numbers are the multiples of their least common multiple.*

This may be readily proved by means of the unique factorization theorem. The method is obvious. We shall, however, give a proof independent of this theorem.

Consider first the case of two numbers; denote them by m and n and their greatest common divisor by d. Then we have

$$m = d\mu, \quad n = d\nu,$$

where μ and ν are relatively prime integers. The common multiples sought are multiples of m and are all comprised in the numbers am, $=ad\mu$, where a is any integer whatever. In order that these numbers shall be multiples of n it is necessary and sufficient that $ad\mu$ shall be a multiple of $d\nu$; that is, that $a\mu$ shall be a multiple of ν; that is, that a shall be a multiple of ν, since μ and ν are relatively prime. Writing $a=\delta\nu$ we have as the multiples in question the set $\delta d\mu\nu$ where δ is an arbitrary integer. This proves the theorem for the case of two numbers; for $d\mu\nu$ is evidently the least common multiple of m and n.

We shall now extend the proposition to any number of integers $m, n, p, q, \ldots$. The multiples in question must be common multiples of m and n and hence of their least common multiple μ. Then the multiples must be multiples of μ and p and hence of their least common multiple μ_1. But μ_1 is evidently the least common multiple of m, n, p. Continuing in a similar manner we may show that every multiple in question is a multiple of μ, the least common multiple of $m, n, p, q, \ldots$. And evidently every such number is a multiple of each of the numbers $m, n, p, q, \ldots$.

Thus the proof of the theorem is complete.

When the two integers m and n are relatively prime their greatest common divisor is 1 and their least common multiple is their product. Again if p is prime to both m and n it is prime to their product mn; and hence the least common multiple of m, n, p is in this case mnp. Continuing in a similar manner we have the theorem:

II. *The least common multiple of several integers, prime each to each, is equal to their product.*

EXERCISES

1. In order that a common multiple of n numbers shall be the least, it is necessary and sufficient that the quotients obtained by dividing it successively by the numbers shall be relatively prime.

2. The product of n numbers is equal to the product of their least common multiple by the greatest common divisor of their products $n-1$ at a time.

3. The least common multiple of n numbers is equal to any common multiple M divided by the greatest common divisor of the quotients obtained on dividing this common multiple by each of the numbers.

4. The product of n numbers is equal to the product of their greatest common divisor by the least common multiple of the products of the numbers taken $n-1$ at a time.

§ 11. Scales of Notation

I. *If m and n are positive integers and* $n>1$, *then m can be represented in terms of n in one and in only one way in the form*

$$m=a_0n^h+a_1n^{h-1}+\ \ldots\ +a_{h-1}n+a_h,$$

where

$$a_0\neq 0,\quad 0\leq a_i<n,\quad i=0,1,2,\ \ldots\ ,h.$$

That such a representation of m exists is readily proved by means of the fundamental theorem of Euclid. For we have

$$\begin{aligned}
m&=n_0n+a_h, & 0&\leq a_h<n,\\
n_0&=n_1n+a_{h-1}, & 0&\leq a_{h-1}<n,\\
n_1&=n_2n+a_{h-2}, & 0&\leq a_{h-2}<n,\\
&\ldots\ldots\ldots\ldots\ldots\ldots\\
n_{h-3}&=n_{h-2}n+a_2, & 0&\leq a_2<n,\\
n_{h-2}&=n_{h-1}n+a_1, & 0&\leq a_1<n,\\
n_{h-1}&=a_0, & 0&<a_0<n.
\end{aligned}$$

If the value of n_{h-1} given in the last of these equations is substituted in the second last we have

$$n_{h-2}=a_0n+a_1.$$

This with the preceding gives

$$n_{h-3}=a_0n^2+a_1n+a_2.$$

Substituting from this in the preceding and continuing the process we have finally

$$m=a_0n^h+a_1n^{h-1}+\ \ldots\ +a_{h-1}n+a_h,$$

a representation of m in the form specified in the theorem.

To prove that this representation is unique, we shall suppose that m has the representation

$$m=b_0n^k+b_1n^{k-1}+\ \ldots\ +b_{k-1}n+b_k,$$

where

$$b_0\neq 0,\quad 0\leq b_i<n,\quad i=0,1,2,\ \ldots\ ,k,$$

and show that the two representations are identical. We have

$$a_0n^h + \ldots + a_{h-1}n + a_h = b_0n^k + \ldots + b_{k-1}n + b_k.$$

Then

$$a_0n^h + \ldots + a_{h-1}n - (b_0n^k + \ldots + b_{k-1}n) = b_k - a_h.$$

The first member is divisible by n. Hence the second is also. But the second member is less than n in absolute value; and hence, in order to be divisible by n, it must be zero. That is, $b_k = a_h$. Dividing the equation through by n and transposing we have

$$a_0n^{h-1} + \ldots + a_{h-2}n - (b_0n^{k-1} + \ldots + b_{k-2}n) = b_{k-1} - a_{h-1}$$

It may now be seen that $b_{k-1} = a_{h-1}$. It is evident that this process may be continued until either the a's are all eliminated from the equation or the b's are all eliminated. But it is obvious that when one of these sets is eliminated the other is also. Hence, $h = k$. Also, every a equals the b which multiplies the same power of n as the corresponding a. That is, the two representations of m are identical. Hence the representation in the theorem is unique.

From this theorem it follows as a special case that any positive integer can be represented in one and in only one way in the scale of 10; that is, in the familiar Hindoo notation. It can also be represented in one and in only one way in any other scale. Thus

$$120759 = 1.7^6 + 0.7^5 + 1.7^4 + 2.7^3 + 0.7^2 + 3.7^1 + 2.$$

Or, using a subscript to denote the scale of notation, this may be written

$$(120759)_{10} = (1012032)_7.$$

For the case in which n (of theorem I) is equal to 2, the only possible values for the a's are 0 and 1. Hence we have at once the following theorem:

II. *Any positive integer can be represented in one and in only one way as a sum of different powers of* 2.

EXERCISES

1. Any positive integer can be represented as an aggregate of different powers of 3, the terms in the aggregate being combined by the signs + and − appropriately chosen.

2. Let m and n be two positive integers of which n is the smaller and suppose that $2^k \leqq n < 2^{k+1}$. By means of the representation of m and n in the scale of 2 prove that the number of divisions to be effected in finding the greatest common divisor of m and n by the Euclidian algorithm does not exceed $2k$.

§ 12. Highest Power of a Prime p Contained in $n!$.

Let n be any positive integer and p any prime number not greater than n. We inquire as to what is the highest power p^ν of the prime p contained in $n!$.

In solving this problem we shall find it convenient to employ the notation

$$\left[\frac{r}{s}\right]$$

to denote the greatest integer α such that $\alpha s \leqq r$. With this notation it is evident that we have

$$\left[\frac{\left[\frac{n}{p}\right]}{p}\right] = \left[\frac{n}{p^2}\right]; \qquad (1)$$

and more generally

$$\left[\frac{\left[\frac{n}{p^i}\right]}{p^j}\right] = \left[\frac{n}{p^{i+j}}\right].$$

If now we use $H\{x\}$ to denote the index of the highest power of p contained in an integer x, it is clear that we have

$$H\{n!\} = H\left\{p \cdot 2p \cdot 3p \ \ldots \ \left[\frac{n}{p}\right] p\right\},$$

since only multiples of p contain the factor p. Hence

$$H\{n!\} = \left[\frac{n}{p}\right] + H\left\{1 \cdot 2 \ \ldots \ \left[\frac{n}{p}\right]\right\}.$$

Applying the same process to the H-function in the second member and remembering relation (1) it is easy to see that we have

$$H\{n!\} = \left[\frac{n}{p}\right] + H\left\{p \cdot 2p \cdot \ldots \cdot \left[\frac{n}{p^2}\right]p\right\}$$

$$= \left[\frac{n}{p}\right] + \left[\frac{n}{p^2}\right] + H\left\{\cdot 1 \cdot 2 \cdot 3 \ldots \left[\frac{n}{p^2}\right]\right\}.$$

Continuing the process we have finally

$$H\{n!\} = \left[\frac{n}{p}\right] + \left[\frac{n}{p^2}\right] + \left[\frac{n}{p^3}\right] + \ldots,$$

the series on the right containing evidently only a finite number of terms different from zero. Thus we have the theorem:

I. *The index of the highest power of a prime p contained in $n!$ is*

$$\left[\frac{n}{p}\right] + \left[\frac{n}{p^2}\right] + \left[\frac{n}{p^3}\right] + \ldots.$$

The theorem just obtained may be written in a different form, more convenient for certain of its applications. Let n be expressed in the scale of p in the form

$$n = a_0 p^h + a_1 p^{h-1} + \ldots + a_{h-1} p + a_h,$$

where

$$a_0 \neq 0, \quad 0 \leqq a_i < p, \quad i = 0, 1, 2, \ldots, h.$$

Then evidently

$$\left[\frac{n}{p}\right] = a_0 p^{h-1} + a_1 p^{h-2} + \ldots + a_{h-2} p + a_{h-1},$$

$$\left[\frac{n}{p^2}\right] = a_0 p^{h-2} + a_1 p^{h-3} + \ldots + a_{h-2},$$

. . . .

Adding these equations member by member and combining the second members in columns as written, we have

$$\left[\frac{n}{p}\right]+\left[\frac{n}{p^2}\right]+\left[\frac{n}{p^3}\right]+\ .\ .\ .$$
$$=\sum_{i=0}^{h}\frac{a_i(p^{h-i}-1)}{p-1}$$
$$=\frac{a_0p^h+a_1p^{h-1}+\ .\ .\ .\ +a_h-(a_0+a_1+\ .\ .\ .\ +a_h)}{p-1}$$
$$=\frac{n-(a_0+a_1+\ .\ .\ .\ +a_h)}{p-1}.$$

Comparing this result with theorem I we have the following theorem:

II. *If n is represented in the scale of p in the form*

$$n=a_0p^h+a_1p^{h-1}+\ .\ .\ .\ +a_h,$$

where p is prime and

$$a_0\neq 0,\quad 0\leqq a_i<p,\quad i=0,\ 1,\ 2,\ .\ .\ .,\ h,$$

then the index of the highest power of p contained in $n!$ is

$$\frac{n-(a_0+a_1+\ .\ .\ .\ +a_h)}{p-1}.$$

Note the simple form of the theorem for the case $p=2$; in this case the denominator $p-1$ is unity.

We shall make a single application of these theorems by proving the following theorem:

III. *If n, α, β, . . . , λ are any positive integers such that $n=\alpha+\beta+\ .\ .\ .\ +\lambda$, then*

$$\frac{n!}{\alpha!\beta!\ .\ .\ .\ \lambda!} \tag{A}$$

is an integer.

Let p be any prime factor of the denominator of the fraction (A). To prove the theorem it is sufficient to show that the index of the highest power of p contained in the numerator is at least as great as the index of the highest power of p contained in the denominator. This index for the denominator is the sum of the expressions

$$\left.\begin{array}{l}\left[\frac{\alpha}{p}\right]+\left[\frac{\alpha}{p^2}\right]+\left[\frac{\alpha}{p^3}\right]+\ldots\\[2mm]\left[\frac{\beta}{p}\right]+\left[\frac{\beta}{p^2}\right]+\left[\frac{\beta}{p^3}\right]+\ldots\\[2mm]\cdot\quad\cdot\quad\cdot\quad\cdot\quad\cdot\quad\cdot\quad\cdot\quad\cdot\quad\cdot\quad\cdot\\[2mm]\left[\frac{\lambda}{p}\right]+\left[\frac{\lambda}{p^2}\right]+\left[\frac{\lambda}{p^3}\right]+\ldots\end{array}\right\}\qquad(B)$$

The corresponding index for the numerator is

$$\left[\frac{n}{p}\right]+\left[\frac{n}{p^2}\right]+\left[\frac{n}{p^3}\right]+\ldots\qquad(C)$$

But, since $n=\alpha+\beta+\ \ldots\ +\lambda$, it is evident that

$$\left[\frac{n}{p^r}\right]\geqq\left[\frac{\alpha}{p^r}\right]+\left[\frac{\beta}{p^r}\right]+\ \ldots\ +\left[\frac{\lambda}{p^r}\right].$$

From this and the expressions in (B) and (C) it follows that the index of the highest power of any prime p in the numerator of (A) is equal to or greater than the index of the highest power of p contained in its denominator. The theorem now follows at once.

COROLLARY. *The product of n consecutive integers is divisible by $n!$.*

EXERCISES

1. Show that the highest power of 2 contained in 1000! is 2^{994}; in 1900! is 2^{1893}. Show that the highest power of 7 contained in 10000! is 7^{1665}.

2. Find the highest power of 72 contained in 1000!

3. Show that 1000! ends with 249 zeros.

4. Show that there is no number n such that 3^7 is the highest power of 3 contained in $n!$.

5. Find the smallest number n such that the highest power of 5 contained in $n!$ is 5^{31}. What other numbers have the same property?

6. If $n=rs$, r and s being positive integers, show that $n!$ is divisible by $(r!)^s$; by $(s!)^r$; by the least common multiple of $(r!)^s$ and $(s!)^r$.

7. If $n=\alpha+\beta+pq+rs$, where α, β, p, q, r, s, are positive integers, then $n!$ is divisible by

$$\alpha!\beta!(q!)^p(s!)^r.$$

8. When m and n are two relatively prime positive integers the quotient

$$Q=\frac{(m+n-1)!}{m!n!}$$

as an integer.

9*. If m and n are positive integers, then each of the quotients

$$Q=\frac{(mn)!}{n!(m!)^n}, \qquad Q=\frac{(2m)!(2n)!}{m!n!(m+n)!},$$

is an integer. Generalize to k integers $m, n, p, \ldots$.

10*. If $n=\alpha+\beta+pq+rs$ where $\alpha, \beta, p, q, r, s$ are positive integers, then $n!$ is divisible by

$$\alpha!\,\beta!\,r!\,p!\,(q!)^p(s!)^r.$$

11*. Show that

$$\frac{(rst)!}{t!(s!)^t(r!)^{st}},$$

is an integer (r, s, t being positive integers). Generalize to the case of n integers $r, s, t, u, \ldots$.

§ 13. Remarks Concerning Prime Numbers

We have seen that the number of primes is infinite. But the integers which have actually been identified as prime are finite in number. Moreover, the question as to whether a large number, as for instance $2^{257}-1$, is prime is in general very difficult to answer. Among the large primes actually identified as such are the following:

$$2^{61}-1, \quad 2^{75}\cdot 5+1, \quad 2^{89}-1, \quad 2^{127}-1.$$

No analytical expression for the representation of prime numbers has yet been discovered. Fermat believed, though he confessed that he was unable to prove, that he had found such an analytical expression in

$$2^{2^n}+1.$$

Euler showed the error of this opinion by finding that 641 is a factor of this number for the case when $n=5$.

The subject of prime numbers is in general one of exceeding difficulty. In fact it is an easy matter to propose problems about prime numbers which no one has been able to solve. Some of the simplest of these are the following:

1. Is there an infinite number of pairs of primes differing by 2?

2. Is every even number (other than 2) the sum of two primes or the sum of a prime and the unit?

3. Is every even number the difference of two primes or the difference of 1 and a prime number?

4. To find a prime number greater than a given prime.

5. To find the prime number which follows a given prime.

6. To find the number of primes not greater than a given number.

7. To compute directly the nth prime number, when n is given.

CHAPTER II

ON THE INDICATOR OF AN INTEGER

§ 14. DEFINITION. INDICATOR OF A PRIME POWER

DEFINITION. If m is any given positive integer the number of positive integers not greater than m and prime to it is called the indicator of m. It is usually denoted by $\phi(m)$, and is sometimes called Euler's ϕ-function of m. More rarely, it has been given the name of totient of m.

As examples we have

$$\phi(1)=1, \quad \phi(2)=1, \quad \phi(3)=2, \quad \phi(4)=2.$$

If p is a prime number it is obvious that

$$\phi(p)=p-1;$$

for each of the integers $1, 2, 3, \ldots, p-1$ is prime to p.

Instead of taking $m=p$ let us assume that $m=p^\alpha$, where α is a positive integer, and seek the value of $\phi(p^\alpha)$. Obviously, every number of the set $1, 2, 3, \ldots, p^\alpha$ either is divisible by p or is prime to p^α. The number of integers in the set divisible by p is $p^{\alpha-1}$. Hence $p^\alpha-p^{\alpha-1}$ of them are prime to p. Hence $\phi(p^\alpha)=p^\alpha-p^{\alpha-1}$. Therefore

If p is any prime number and α is any positive integer, then

$$\phi(p^\alpha)=p^\alpha\left(1-\frac{1}{p}\right).$$

§ 15. THE INDICATOR OF A PRODUCT

I. *If μ and ν are any two relatively prime positive integers, then*

$$\phi(\mu\nu)=\phi(\mu)\,\phi(\nu).$$

In order to prove this theorem let us write all the integers up to $\mu\nu$ in a rectangular array as follows:

$$\left.\begin{matrix} 1 & 2 & 3 \ldots & h \ldots & \mu \\ \mu+1 & \mu+2 & \mu+3 \ldots & \mu+h \ldots & 2\mu \\ 2\mu+1 & 2\mu+2 & 2\mu+3 \ldots & 2\mu+h \ldots & 3\mu \\ \cdot & \cdot & \cdot & \cdot & \cdot \\ (\nu-1)\mu+1 & (\nu-1)\mu+2 & (\nu-1)\mu+3 \ldots & (\nu-1)\mu+h \ldots & \nu\mu \end{matrix}\right\} (A)$$

If a number h in the first line of this array has a factor in common with μ then every number in the same column with h has a factor in common with μ. On the other hand if h is prime to μ so is every number in the column with h at the top. But the number of integers in the first row prime to μ is $\phi(\mu)$. Hence the number of columns containing integers prime to μ is $\phi(\mu)$ and every integer in these columns is prime to μ.

Let us now consider what numbers in one of these columns are prime to ν; for instance, the column with h at the top. We wish to determine how many integers of the set

$$h, \quad \mu+h, \quad 2\mu+h, \quad \ldots, \quad (\nu-1)\mu+h$$

are prime to ν. Write

$$s\mu+h=q_s\nu+r_s$$

where s ranges over the numbers $s=0, 1, 2, \ldots, \nu-1$ and $0\leqq r_s<\nu$. Clearly $s\mu+h$ is or is not prime to ν according as r_s is or is not prime to ν. Our problem is then reduced to that of determining how many of the quantities r_s are prime to ν.

First let us notice that all the numbers r_s are different; for, if $r_s=r_t$ then from

$$s\mu+h=q_s\nu+r_s, \quad t\mu+h=q_t\nu+r_t,$$

we have by subtraction that $(s-t)\mu$ is divisible by ν. But μ is prime to ν and s and t are each less than ν. Hence $(s-t)\mu$ can be a multiple of ν only by being zero; that is, s must equal t. Hence no two of the remainders r_s can be equal.

Now the remainders r_s are ν in number, are all zero or positive, each is less than ν, and they are all distinct. Hence they are in some order the numbers 0, 1, 2, . . . , $\nu-1$. The number of integers in this set prime to ν is evidently $\phi(\nu)$.

Hence it follows that in any column of the array (A) in which the numbers are prime to μ there are just $\phi(\nu)$ numbers which are prime to ν. That is, in this column there are just $\phi(\nu)$ numbers which are prime to $\mu\nu$. But there are $\phi(\mu)$ such columns. Hence the number of integers in the array (A) prime to $\mu\nu$ is $\phi(\mu)\phi(\nu)$.

But from the definition of the ϕ-function it follows that the number of integers in the array (A) prime to $\mu\nu$ is $\phi(\mu\nu)$. Hence,

$$\phi(\mu\nu) = \phi(\mu)\,\phi(\nu),$$

which is the theorem to be proved.

COROLLARY. *In the series of n consecutive terms of an arithmetical progression the common difference of which is prime to n, the number of terms prime to n is $\phi(n)$.*

From theorem I we have readily the following more general result:

II. *If $m_1, m_2, \ldots, m_k$ are k positive integers which are prime each to each, then*

$$\phi(m_1 m_2 \ldots m_k) = \phi(m_1)\,\phi(m_2) \ldots \phi(m_k).$$

§ 16. THE INDICATOR OF ANY POSITIVE INTEGER

From the results of §§ 14 and 15 we have an immediate proof of the following fundamental theorem:

If $m = p_1^{\alpha_1} p_2^{\alpha_2} \ldots p_n^{\alpha_n}$ where $p_1, p_2, \ldots, p_n$ are different primes and $\alpha_1, \alpha_2, \ldots, \alpha_n$ are positive integers, then

$$\phi(m) = m\left(1-\frac{1}{p_1}\right)\left(1-\frac{1}{p_2}\right) \ldots \left(1-\frac{1}{p_n}\right).$$

For,

$$\begin{aligned}\phi(m) &= \phi(p_1^{\alpha_1})\,\phi(p_2^{\alpha_2}) \ldots \phi(p_n^{\alpha_n})\\ &= p_1^{\alpha_1}\left(1-\frac{1}{p_1}\right)p_2^{\alpha_2}\left(1-\frac{1}{p_2}\right) \ldots p_n^{\alpha_n}\left(1-\frac{1}{p_n}\right)\\ &= m\left(1-\frac{1}{p_1}\right)\left(1-\frac{1}{p_2}\right) \ldots \left(1-\frac{1}{p_n}\right).\end{aligned}$$

On account of the great importance of this theorem we shall give a second demonstration of it.

It is clear that the number of integers less than m and divisible by p_1 is

$$\frac{m}{p_1}.$$

The number of integers less than m and divisible by p_2 is

$$\frac{m}{p_2}.$$

The number of integers less than m and divisible by p_1p_2 is

$$\frac{m}{p_1p_2}.$$

Hence the number of integers less than m and divisible by either p_1 or p_2 is

$$\frac{m}{p_1}+\frac{m}{p_2}-\frac{m}{p_1p_2}.$$

Hence the number of integers less than m and prime to p_1p_2 is

$$m-\frac{m}{p_1}-\frac{m}{p_2}+\frac{m}{p_1p_2}=m\left(1-\frac{1}{p_1}\right)\left(1-\frac{1}{p_2}\right).$$

We shall now show that if the number of integers less than m and prime to $p_1p_2\ .\ .\ .\ p_i$, where i is less than n, is

$$m\left(1-\frac{1}{p_1}\right)\left(1-\frac{1}{p_2}\right)\ .\ .\ .\ \left(1-\frac{1}{p_i}\right),$$

then the number of integers less than m and prime to $p_1p_2\ .\ .\ .\ p_ip_{i+1}$ is

$$m\left(1-\frac{1}{p_1}\right)\left(1-\frac{1}{p_2}\right)\ .\ .\ .\ \left(1-\frac{1}{p_{i+1}}\right).$$

From this our theorem will follow at once by induction.

From our hypothesis it follows that the number of integers less than m and divisible by at least one of the primes $p_1, p_2, \ldots, p_i$ is

$$m - m\left(1-\frac{1}{p_1}\right) \ldots \left(1-\frac{1}{p_i}\right),$$

or

$$\Sigma\frac{m}{p_1} - \Sigma\frac{m}{p_1p_2} + \Sigma\frac{m}{p_1p_2p_3} - \ldots, \qquad (A)$$

where the summation in each case runs over all numbers of the type indicated, the subscripts of the p's being equal to or less than i.

Let us consider the integers less than m and having the factor p_{i+1} but not having any of the factors $p_1, p_2, \ldots, p_i$. Their number is

$$\frac{m}{p_{i+1}} - \frac{1}{p_{i+1}}\left\{\Sigma\frac{m}{p_1} - \Sigma\frac{m}{p_1p_2} + \Sigma\frac{m}{p_1p_2p_3} - \ldots\right\}, \qquad (B)$$

where the summation signs have the same significance as before. For the number in question is evidently m/p_{i+1} *minus* the number of integers not greater than m/p_{i+1} and divisible by at least one of the primes $p_1, p_2, \ldots, p_i$.

If we add (A) and (B) we have the number of integers less than m and divisible by one at least of the numbers $p_1, p_2, \ldots, p_{i+1}$. Hence the number of integers less than m and prime to $p_1, p_2, \ldots, p_{i+1}$ is

$$m - \Sigma\frac{m}{p_1} + \Sigma\frac{m}{p_1p_2} - \Sigma\frac{m}{p_1p_2p_3} + \ldots,$$

where now in the summations the subscripts run from 1 to $i+1$. This number is clearly equal to

$$m\left(1-\frac{1}{p_1}\right)\left(1-\frac{1}{p_2}\right) \ldots \left(1-\frac{1}{p_{i+1}}\right).$$

From this result, as we have seen above, our theorem follows at once by induction.

§ 17. Sum of the Indicators of the Divisors of a Number

We shall first prove the following lemma:

Lemma. If d is any divisor of m and $m = nd$, the number of integers not greater than m which have with m the greatest common divisor d is $\phi(n)$.

Every integer not greater than m and having the divisor d is contained in the set $d, 2d, 3d, \ldots, nd$. The number of these integers which have with m the greatest common divisor d is evidently the same as the number of integers of the set $1, 2, \ldots, n$ which are prime to m/d, or n; for αd and m have or have not the greatest common divisor d according as α is or is not prime to m/d, $=n$. Hence the number in question is $\phi(n)$.

From this lemma follows readily the proof of the following theorem:

If $d_1, d_2, \ldots, d_r$ are the different divisors of m, then

$$\phi(d_1) + \phi(d_2) + \ldots + \phi(d_r) = m.$$

Let us define integers $m_1, m_2, \ldots, m_r$ by the relations

$$m = d_1 m_1 = d_2 m_2 = \ldots = d_r m_r.$$

Now consider the set of m positive integers not greater than m, and classify them as follows into r classes. Place in the first class those integers of the set which have with m the greatest common divisor m_1; their number is $\phi(d_1)$, as may be seen from the lemma. Place in the second class those integers of the set which have with m the greatest common divisor m_2; their number is $\phi(d_2)$. Proceeding in this way throughout, we place finally in the last class those integers of the set which have with m the greatest common divisor m_r; their number is $\phi(d_r)$. It is evident that every integer in the set falls into one and into just one of these r classes. Hence the total number m of integers in the set is $\phi(d_1) + \phi(d_2) + \ldots + \phi(d_r)$. From this the theorem follows immediately.

EXERCISES

1. Show that the indicator of any integer greater than 2 is even.

2. Prove that the number of irreducible fractions not greater than 1 and with denominator equal to n is $\phi(n)$.

3. Prove that the number of irreducible fractions not greater than 1 and with denominators not greater than n is

$$\phi(1)+\phi(2)+\phi(3)+\ \ldots\ +\phi(n).$$

4. Show that the sum of the integers less than n and prime to n is $\frac{1}{2}n\phi(n)$ if $n>1$.

5. Find ten values of x such that $\phi(x)=24$.

6. Find seventeen values of x such that $\phi(x)=72$.

7. Find three values of n for which there is no x satisfying the equation $\phi(x)=2n$.

8. Show that if the equation

$$\phi(x)=n$$

has one solution it always has a second solution, n being given and x being the unknown.

9. Prove that all the solutions of the equation

$$\phi(x)=4n-2, \quad n>1,$$

are of the form p^α and $2p^\alpha$, where p is a prime of the form $4s-1$.

10. How many integers prime to n are there in the set

a) $1\cdot 2,\quad 2\cdot 3,\quad 3\cdot 4,\ \ldots,\quad n(n+1)$?

b) $1\cdot 2\cdot 3,\quad 2\cdot 3\cdot 4,\quad 3\cdot 4\cdot 5,\ \ldots,\quad n(n+1)(n+2)$?

c) $\frac{1\cdot 2}{2},\quad \frac{2\cdot 3}{2},\quad \frac{3\cdot 4}{2},\ \ldots,\quad \frac{n(n+1)}{2}$?

d) $\frac{1\cdot 2\cdot 3}{6},\quad \frac{2\cdot 3\cdot 4}{6},\quad \frac{3\cdot 4\cdot 5}{6},\ \ldots,\quad \frac{n(n+1)(n+2)}{6}$?

11*. Find a method for determining all the solutions of the equation

$$\phi(x)=n,$$

where n is given and x is to be sought.

12*. A number theory function $\phi(n)$ is defined for every positive integer n; and for every such number n it satisfies the relation

$$\phi(d_1)+\phi(d_2)+\ \ldots\ +\phi(d_r)=n,$$

where $d_1, d_2, \ldots, d_r$ are the divisors of n. From this property alone show that

$$\phi(n)=n\left(1-\frac{1}{p_1}\right)\left(1-\frac{1}{p_2}\right)\ \ldots\ \left(1-\frac{1}{p_k}\right),$$

where $p_1, p_2, \ldots, p_k$ are the different prime factors of n.

CHAPTER III

ELEMENTARY PROPERTIES OF CONGRUENCES

§ 18. Congruences Modulo m

Definitions. If a and b are any two integers, positive or zero or negative, whose difference is divisible by m, a and b are said to be congruent modulo m, or congruent for the modulus m, or congruent according to the modulus m. Each of the numbers a and b is said to be a residue of the other.

To express the relation thus defined we may write

$$a = b + cm,$$

where c is an integer (positive or zero or negative). It is more convenient, however, to use a special notation due to Gauss, and to write

$$a \equiv b \bmod m,$$

an expression which is read a is congruent to b modulo m, or a is congruent to b for the modulus m, or a is congruent to b according to the modulus m. This notation has the advantage that it involves only the quantities which are essential to the idea involved, whereas in the preceding expression we had the irrelevant integer c. The Gaussian notation is of great value and convenience in the study of the theory of divisibility. In the present chapter we develop some of the fundamental elementary properties of congruences. It will be seen that many theorems concerning equations are likewise true of congruences with fixed modulus; and it is this analogy with equations which gives congruences (as such) one of their chief claims to attention.

As immediate consequences of our definitions we have the following fundamental theorems:

I. *If* $$a \equiv c \bmod m, \quad b \equiv c \bmod m,$$

then $$a \equiv b \bmod m;$$

that is, for a given modulus, numbers congruent to the same number are congruent to each other.

For, by hypothesis, $a-c=c_1m$, $b-c=c_2m$, where c_1 and c_2 are integers. Then by subtraction we have $a-b=(c_1-c_2)m$; whence $a \equiv b \bmod m$.

II. *If* $$a \equiv b \bmod m, \quad \alpha \equiv \beta \bmod m,$$

then $$a \pm \alpha \equiv b \pm \beta \bmod m;$$

that is, congruences with the same modulus may be added or subtracted member by member.

For, by hypothesis, $a-b=c_1m$, $\alpha-\beta=c_2m$; whence $(a \pm \alpha)-(b \pm \beta)=(c_1 \pm c_2)m$. Hence $a \pm \alpha \equiv b \pm \beta \bmod m$.

III. *If* $$a \equiv b \bmod m,$$

then $$ca \equiv cb \bmod m,$$

c being any integer whatever.

The proof is obvious and need not be stated.

IV. *If* $$a \equiv b \bmod m, \quad \alpha \equiv \beta \bmod m,$$

then $$a\alpha \equiv b\beta \bmod m;$$

that is, two congruences with the same modulus may be multiplied member by member.

For, we have $a=b+c_1m$, $\alpha=\beta+c_2m$. Multiplying these equations member by member we have $a\alpha=b\beta+m(bc_2+\beta c_1+c_1c_2m)$. Hence $a\alpha \equiv b\beta \bmod m$.

A repeated use of this theorem gives the following result:

V. *If* $$a \equiv b \bmod m,$$

then $$a^n \equiv b^n \bmod m$$

where n is any positive integer.

As a corollary of theorems II, III and V we have the following more general result:

VI. *If* $f(x)$ *denotes any polynomial in* x *with coefficients which are integers (positive or zero or negative) and if further* $a \equiv b \mod m$, *then*

$$f(a) \equiv f(b) \mod m.$$

§ 19. Solutions of Congruences by Trial

Let $f(x)$ be any polynomial in x with coefficients which are integers (positive or negative or zero). Then if x and c are any two integers it follows from the last theorem of the preceding section that

$$f(x) \equiv f(x+cm) \mod m. \qquad (1)$$

Hence if α is any value of x for which the congruence

$$f(x) \equiv 0 \mod m \qquad (2)$$

is satisfied, then the congruence is also satisfied for $x=\alpha+cm$, where c is any integer whatever. The numbers $\alpha+cm$ are said to form a *solution* (or to be a *root*) of the congruence, c being a variable integer. Any one of the integers $\alpha+cm$ may be taken as the representative of the solution. We shall often speak of *one* of these numbers as the solution itself.

Among the integers in a solution of the congruence (2) there is evidently one which is positive and not greater than m. Hence all solutions of a congruence of the type (2) may be found by trial, a substitution of each of the numbers 1, 2, . . . , m being made for x. It is clear also that m is the maximum number of solutions which (2) can have whatever be the function $f(x)$. By means of an example it is easy to show that this maximum number of solutions is not always possessed by a congruence; in fact, it is not even necessary that the congruence have a solution at all.

This is illustrated by the example

$$x^2-3 \equiv 0 \mod 5.$$

In order to show that no solution is possible it is necessary to make trial only of the values 1, 2, 3, 4, 5 for x. A direct substitution verifies the conclusion that none of them satisfies the congruence; and hence that the congruence has no solution at all.

On the other hand the congruence

$$x^5 - x \equiv 0 \bmod 5$$

has the solutions $x = 1, 2, 3, 4, 5$ as one readily verifies; that is, this congruence has five solutions—the maximum number possible in accordance with the results obtained above.

EXERCISES

1. Show that

$$(a+b)^p \equiv a^p + b^p \bmod p$$

where a and b are any integers and p is any prime.

2. From the preceding result prove that

$$\alpha^p \equiv \alpha \bmod p$$

for every integer α.

3. Find all the solutions of each of the congruences $x^{11} \equiv x \bmod 11$, $x^{10} \equiv 1 \bmod 11$, $x^5 \equiv 1 \bmod 11$.

§ 20. Properties of Congruences Relative to Division

The properties of congruences relative to addition, subtraction and multiplication are entirely analogous to the properties of algebraic equations. But the properties relative to division are essentially different. These we shall now give.

I. *If two numbers are congruent modulo m they are congruent modulo d, where d is any divisor of m.*

For, from $a \equiv b \bmod m$, we have $a = b + cm = b + c'd$. Hence $a \equiv b \bmod d$.

II. *If two numbers are congruent for different moduli they are congruent for a modulus which is the least common multiple of the given moduli.*

The proof is obvious, since the difference of the given numbers is divisible by each of the moduli.

III. *When the two members of a congruence are multiples of an integer c prime to the modulus, each member of the congruence may be divided by c.*

For, if $ca \equiv cb \bmod m$ then $ca - cb$ is divisible by m. Since c is prime to m it follows that $a - b$ is divisible by m. Hence $a \equiv b \bmod m$.

IV. *If the two members of a congruence are divisible by an integer c, having with the modulus the greatest common divisor δ, one obtains a congruence equivalent to the given congruence by dividing the two members by c and the modulus by δ.*

By hypothesis

$$ac \equiv bc \bmod m, \quad c = \delta c_1, \quad m = \delta m_1.$$

Hence $c(a-b)$ is divisible by m. A necessary and sufficient condition for this is evidently that $c_1(a-b)$ is divisible by m_1. This leads at once to the desired result.

§ 21. Congruences with a Prime Modulus

The congruence *

$$a_0x^n + a_1x^{n-1} + \ . \ . \ . \ + a_n \equiv 0 \bmod p, \quad a_0 \not\equiv 0 \bmod p,$$

where p is a prime number and the a's are any integers, has not more than n solutions.

Denote the first member of this congruence by $f(x)$ so that the congruence may be written

$$f(x) \equiv 0 \bmod p. \qquad (1)$$

Suppose that a is a root of the congruence, so that

$$f(a) \equiv 0 \bmod p.$$

Then we have

$$f(x) \equiv f(x) - f(a) \bmod p.$$

* The sign $\not\equiv$ is read *is not congruent to.*

But, from algebra, $f(x)-f(a)$ is divisible by $x-a$. Let $(x-a)^\alpha$ be the highest power of $x-a$ contained in $f(x)-f(a)$. Then we may write

$$f(x)-f(a)=(x-a)^\alpha f_1(x), \tag{2}$$

where $f_1(x)$ is evidently a polynomial with integral coefficients. Hence we have

$$f(x)\equiv(x-a)^\alpha f_1(x) \bmod p. \tag{3}$$

We shall say that a occurs α times as a solution of (1); or that the congruence has α solutions each equal to a.

Now suppose that congruence (1) has a root b such that $b\not\equiv a \bmod p$. Then from (3) we have

$$f(b)\equiv(b-a)^\alpha f_1(b) \bmod p.$$

But $\qquad f(b)\equiv 0 \bmod p, \quad (b-a)^\alpha\not\equiv 0 \bmod p.$

Hence, since p is a prime number, we must have

$$f_1(b)\equiv 0 \bmod p.$$

By an argument similar to that just used above, we may show that $f_1(x)-f_1(b)$ may be written in the form

$$f_1(x)-f_1(b)=(x-b)^\beta f_2(x),$$

where β is some positive integer. Then we have

$$f(x)\equiv(x-a)^\alpha(x-b)^\beta f_2(x) \bmod p.$$

Now this process can be continued until either all the solutions of (1) are exhausted or the second member is a product of linear factors multiplied by the integer a_0. In the former case there will be fewer than n solutions of (1), so that our theorem is true for this case. In the other case we have

$$f(x)\equiv a_0(x-a)^\alpha(x-b)^\beta \ . \ . \ . \ (x-l)^\lambda \bmod p.$$

We have now n solutions of (1): a counted α times, b counted β times, . . . , l counted λ times; $\alpha+\beta+ \ . \ . \ . \ +\lambda=n$.

Now let η be any solution of (1). Then

$$f(\eta) \equiv a_0(\eta-a)^\alpha(\eta-b)^\beta \ . \ . \ . \ (\eta-l)^\lambda \equiv 0 \bmod p.$$

Since p is prime it follows now that some one of the factors $\eta-a$, $\eta-b$, . . . , $\eta-l$ is divisible by p. Hence η coincides with one of the solutions a, b, c, . . . , l. That is, (1) can have only the n solutions already found.

This completes the proof of the theorem.

EXERCISES

1. Construct a congruence of the form

$$a_0x^n + a_1x^{n-1} + . \ . \ . \ + a_n \equiv 0 \bmod m, \quad a_0 \not\equiv 0 \bmod m,$$

having more than n solutions and thus show that the limitation to a prime modulus in the theorem of this section is essential.

2. Prove that

$$x^6 - 1 \equiv (x-1)(x-2)(x-3)(x-4)(x-5)(x-6) \bmod 7$$

for every integer x.

3. How many solutions has the congruence $x^5 \equiv 1 \bmod 11$? the congruence $x^5 \equiv 2 \bmod 11$?

§ 22. LINEAR CONGRUENCES

From the theorem of the preceding section it follows that the congruence

$$ax \equiv c \bmod p, \quad a \not\equiv 0 \bmod p,$$

where p is a prime number, has not more than one solution. In this section we shall prove that it always has a solution. More generally, we shall consider the congruence

$$ax \equiv c \bmod m$$

where m is any integer. The discussion will be broken up into parts for convenience in the proofs.

I. *The congruence*

$$ax \equiv 1 \bmod m, \qquad (1)$$

in which a and m are relatively prime, has one and only one solution.

The question as to the existence and number of the solutions of (1) is equivalent to the question as to the existence and number of integer pairs x, y satisfying the equation,

$$ax - my = 1, \tag{2}$$

the integers x being incongruent modulo m. Since a and m are relatively prime it follows from theorem IV of § 9 that there exists a solution of equation (2). Let $x = \alpha$ and $y = \beta$ be a particular solution of (2) and let $x = \bar{\alpha}$ and $y = \bar{\beta}$ be any solution of (2). Then we have

$$a\alpha - m\beta = 1,$$
$$a\bar{\alpha} - m\bar{\beta} = 1;$$

whence

$$a(\alpha - \bar{\alpha}) - m(\beta - \bar{\beta}) = 0.$$

Hence $\alpha - \bar{\alpha}$ is divisible by m, since a and m are relatively prime. That is, $\alpha \equiv \bar{\alpha} \bmod m$. Hence α and $\bar{\alpha}$ are representatives of the same solution of (1). Hence (1) has one and only one solution, as was to be proved.

II. *The solution $x = \alpha$ of the congruence $ax \equiv 1 \bmod m$, in which a and m are relatively prime, is prime to m.*

For, if $a\alpha - 1$ is divisible by m, α is divisible by no factor of m except 1.

III. *The congruence*

$$ax \equiv c \bmod m \tag{3}$$

in which a and m and also c and m are relatively prime, has one and only one solution.

Let $x = \gamma$ be the unique solution of the congruence $cx \equiv 1 \bmod m$. Then we have $a\gamma x \equiv c\gamma \equiv 1 \bmod m$. Now, by I we see that there is one and only one solution of the congruence $a\gamma x \equiv 1 \bmod m$; and from this the theorem follows at once.

Suppose now that a is prime to m but that c and m have the greatest common divisor δ which is different from 1. Then it is easy to see that any solution x of the congruence $ax \equiv c \bmod m$

must be divisible by δ. The question of the existence of solutions of the congruence $ax \equiv c \bmod m$ is then equivalent to the question of the existence of solutions of the congruence

$$a\frac{x}{\delta} \equiv \frac{c}{\delta} \bmod \frac{m}{\delta},$$

where x/δ is the unknown integer. From III it follows that this congruence has a unique solution $x/\delta = \alpha$. Hence the congruence $ax \equiv c \bmod m$ has the unique solution $x = \delta\alpha$. Thus we have the following theorem:

IV. *The congruence* $ax \equiv c \bmod m$, *in which* a *and* m *are relatively prime, has one and only one solution.*

COROLLARY. The congruence $ax \equiv c \bmod p$, $a \not\equiv 0 \bmod p$, *where* p *is a prime number, has one and only one solution.*

It remains to examine the case of the congruence $ax \equiv c \bmod m$ in which a and m have the greatest common divisor d. It is evident that there is no solution unless c also contains this divisor d. Then let us suppose that $a = \alpha d$, $c = \gamma d$, $m = \mu d$. Then for every x such that $ax \equiv c \bmod m$ we have $\alpha x \equiv \gamma \bmod \mu$; and conversely every x satisfying the latter congruence also satisfies the former. Now $\alpha x \equiv \gamma \bmod \mu$ has only one solution. Let β be a non-negative number less than μ which satisfies the congruence $\alpha x \equiv \gamma \bmod \mu$. All integers which satisfy this congruence are then of the form $\beta + \mu\nu$, where ν is an integer. Hence all integers satisfying the congruence $ax \equiv c \bmod m$ are of the form $\beta + \mu\nu$; and every such integer is a representative of a solution of this congruence. It is clear that the numbers

$$\beta, \quad \beta + \mu, \quad \beta + 2\mu, \quad \ldots, \quad \beta + (d-1)\mu \qquad (A)$$

are incongruent modulo m while every integer of the form $\beta + \mu\nu$ is congruent modulo m to a number of the set (A). Hence the congruence $ax \equiv c \bmod m$ has the d solutions (A).

This leads us to an important theorem which includes all the other theorems of this section as special cases. It may be stated as follows:

V. *Let*

$$ax \equiv c \bmod m$$

be any linear congruence and let a and m have the greatest common divisor d $(d \geq 1)$. *Then a necessary and sufficient condition for the existence of solutions of the congruence is that c be divisible by d. If this condition is satisfied the congruence has just d solutions, and all the solutions are congruent modulo m/d.*

EXERCISES

1. Find the remainder when 2^{40} is divided by 31; when 2^{43} is divided by 31.

2. Show that $2^{2^5}+1$ has the factor 641.

3. Prove that a number is a multiple of 9 if and only if the sum of its digits is a multiple of 9.

4. Prove that a number is a multiple of 11 if and only if the sum of the digits in the odd numbered places diminished by the sum of the digits in the even numbered places is a multiple of 11.

CHAPTER IV

THE THEOREMS OF FERMAT AND WILSON

§ 23. Fermat's General Theorem

Let m be any positive integer and let

$$a_1, \quad a_2, \quad . \ . \ . \ , \quad a_{\phi(m)} \qquad (A)$$

be the set of $\phi(m)$ positive integers not greater than m and prime to m. Let a be any integer prime to m and form the set of integers

$$aa_1, \quad aa_2, \quad . \ . \ . \ , \quad aa_{\phi(m)}. \qquad (B)$$

No number aa_i of the set (B) is congruent to a number aa_j unless $j=i$; for, from

$$aa_i \equiv aa_j \bmod m$$

we have $a_i \equiv a_j \bmod m$; whence $a_i = a_j$, since both a_i and a_j are positive and not greater than m. Therefore $j=i$. Furthermore, every number of the set (B) is congruent to some number of the set (A). Hence we have congruences of the form

$$aa_1 \equiv a_{i_1} \bmod m,$$
$$aa_2 \equiv a_{i_2} \bmod m,$$
$$. \ . \ . \ . \ . \ . \ . \ .$$
$$aa_{\phi(m)} \equiv a_{i_{\phi(m)}} \bmod m.$$

No two numbers in the second members are equal, since $aa_i \not\equiv aa_j$ unless $i=j$. Hence the numbers a_{i_1}, a_{i_2}, . . . , $a_{i_{\phi(m)}}$ are the numbers a_1, a_2, . . . , $a_{\phi(m)}$ in some order. Therefore, if we multiply the above system of congruences together member by member and divide each member of the resulting congruence by $a_1 \cdot a_2 \ldots a_{\phi(m)}$ (which is prime to m), we have

$$a^{\phi(m)} \equiv 1 \bmod m.$$

This result is known as Fermat's general theorem. It may be stated as follows:

If m is any positive integer and a is any integer prime to m, then

$$a^{\phi(m)} \equiv 1 \bmod m.$$

COROLLARY 1. *If a is any integer not divisible by a prime number p, then*

$$a^{p-1} \equiv 1 \bmod p.$$

COROLLARY 2. *If p is any prime number and a is any integer, then*

$$a^p \equiv a \bmod p.$$

§ 24. Euler's Proof of the Simple Fermat Theorem

The theorem of Cor. 1, § 23, is often spoken of as the simple Fermat theorem. It was first announced by Fermat in 1679, but without proof. The first proof of it was given by Euler in 1736. This proof may be stated as follows:

From the Binomial Theorem it follows readily that

$$(a+1)^p \equiv a^p + 1 \bmod p$$

since

$$\frac{p!}{r!(p-r)!}, \quad 0 < r < p,$$

is obviously divisible by p. Subtracting $a+1$ from each side of the foregoing congruence, we have

$$(a+1)^p - (a+1) \equiv a^p - a \bmod p.$$

Hence if a^p-a is divisible by p, so is $(a+1)^p-(a+1)$. But 1^p-1 is divisible by p. Hence 2^p-2 is divisible by p; and then 3^p-3; and so on. Therefore, in general, we have

$$a^p \equiv a \bmod p.$$

If a is prime to p this gives $a^{p-1} \equiv 1 \bmod p$, as was to be proved.

If instead of the Binomial Theorem one employs the Polynomial Theorem, an even simpler proof is obtained. For, from the latter theorem, we have readily

$$(\alpha_1+\alpha_2+ \ . \ . \ . \ +\alpha_a)^p \equiv \alpha_1{}^p+\alpha_2{}^p+ \ . \ . \ . \ +\alpha_a{}^p \bmod p.$$

Putting $\alpha_1=\alpha_2= \ . \ . \ . \ =\alpha_a=1$ we have

$$a^p \equiv a \bmod p,$$

from which the theorem follows as before.

§ 25. Wilson's Theorem

From the simple Fermat theorem it follows that the congruence

$$x^{p-1} \equiv 1 \bmod p$$

has the $p-1$ solutions $1, 2, 3, \ . \ . \ . \ , p-1$. Hence from the discussion in § 21 it follows that

$$x^{p-1}-1 \equiv (x-1)(x-2) \ . \ . \ . \ (x-\overline{p-1}) \bmod p,$$

this relation being satisfied for every value of x. Putting $x=0$ we have

$$(-1) \equiv (-1)^{p-1} \cdot 1 \cdot 2 \cdot 3 \ . \ . \ . \ \overline{p-1} \bmod p.$$

If p is an odd prime this leads to the congruence

$$1 \cdot 2 \cdot 3 \ . \ . \ . \ \overline{p-1}+1 \equiv 0 \bmod p.$$

Now for $p=2$ this congruence is evidently satisfied. Hence we have the Wilson theorem:

Every prime number p satisfies the relation

$$1 \cdot 2 \cdot 3 \ . \ . \ . \ \overline{p-1}+1 \equiv 0 \bmod p.$$

An interesting proof of this theorem on wholly different principles may be given. Let p points be distributed at equal intervals on the circumference of a circle. The whole number of p-gons which can be formed by joining up these p points in every possible order is evidently

$$\frac{1}{2p}p(p-1)(p-2)\ \ldots\ 3\cdot 2\cdot 1;$$ [1]

for the first vertex can be chosen in p ways, the second in $p-1$ ways, . . . , the $(p-1)^{\text{th}}$ in two ways, and the last in one way; and in counting up thus we have evidently counted each polygon $2p$ times, once for each vertex and for each direction from the vertex around the polygon. Of the total number of polygons $\frac{1}{2}(p-1)$ are regular (convex or stellated) so that a revolution through $360°/p$ brings each of these into coincidence with its former position. The number of remaining p-gons must be divisible by p; for with each such p-gon we may associate the $p-1$ p-gons which can be obtained from it by rotating it through successive angles of $360°/p$. That is,

$$\frac{1}{2p}p(p-1)(p-2)\ \ldots\ 3\cdot 2\cdot 1-\frac{1}{2}(p-1)\equiv 0 \bmod p.$$

Hence

$$(p-1)(p-2)\ \ldots\ 3\cdot 2\cdot 1-p+1\equiv 0 \bmod p;$$

and from this it follows that

$$1\cdot 2\ \ldots\ \overline{p-1}+1\equiv 0 \bmod p,$$

as was to be proved.

§ 26. The Converse of Wilson's Theorem

Wilson's theorem is noteworthy in that its converse is also true. The converse may be stated as follows:

Every integer n such that the congruence

$$1\cdot 2\cdot 3\ \ldots\ \overline{n-1}+1\equiv 0 \bmod n$$

is satisfied is a prime number.

For, if n is not prime, there is some divisor d of n different from 1 and less than n. For such a d we have $1 \cdot 2 \cdot 3 \ldots \overline{n-1} \equiv 0 \bmod d$; so that $1 \cdot 2 \ldots \overline{n-1}+1 \not\equiv 0 \bmod d$; and hence $1 \cdot 2 \ldots \overline{n-1}+1 \not\equiv 0 \bmod n$. Since this contradicts our hypothesis the truth of the theorem follows.

Wilson's theorem and its converse may be combined into the following elegant theorem:

A necessary and sufficient condition that an integer n is prime is that

$$1 \cdot 2 \cdot 3 \ldots \overline{n-1}+1 \equiv 0 \bmod n.$$

Theoretically this furnishes a complete and elegant test as to whether a given number is prime. But, practically, the labor of applying it is so great that it is useless for verifying large primes.

§ 27. Impossibility of $1 \cdot 2 \cdot 3 \ldots \overline{n-1}+1=n^k$ for $n>5$.

In this section we shall prove the following theorem:

There exists no integer k for which the equation

$$1 \cdot 2 \cdot 3 \ldots \overline{n-1}+1=n^k$$

is true when n is greater than 5.

If n contains a divisor d different from 1 and n, the equation is obviously false; for the second member is divisible by d while the first is not. Hence we need to prove the theorem only for primes n.

Transposing 1 to the second member and dividing by $n-1$ we have

$$1 \cdot 2 \cdot 3 \ldots \overline{n-2}=n^{k-1}+n^{k-2}+ \ldots +n+1.$$

If $n>5$ the product on the left contains both the factor 2 and the factor $\frac{1}{2}(n-1)$; that is, the first member contains the factor $n-1$. But the second member does not contain this factor, since for $n=1$ the expression $n^{k-1}+ \ldots +n+1$ is equal to $k \neq 0$. Hence the theorem follows at once.

§ 28. Extension of Fermat's Theorem

The object of this section is to extend Fermat's general theorem and incidentally to give a new proof of it. We shall base this proof on the simple Fermat theorem, of which we have already given a simple independent proof. This theorem asserts that for every prime p and integer a not divisible by p, we have the congruence

$$a^{p-1} \equiv 1 \bmod p.$$

Then let us write

$$a^{p-1} = 1 + hp. \tag{1}$$

Raising each member of this equation to the pth power we may write the result in the form

$$a^{p(p-1)} = 1 + h_1 p^2 \tag{2}$$

where h_1 is an integer. Hence

$$a^{p(p-1)} \equiv 1 \bmod p^2.$$

By raising each member of (2) to the pth power we can readily show that

$$a^{p^2(p-1)} \equiv 1 \bmod p^3.$$

It is now easy to see that we shall have in general

$$a^{p^{\alpha-1}(p-1)} \equiv 1 \bmod p^\alpha,$$

where α is a positive integer; that is,

$$a^{\phi(p^\alpha)} \equiv 1 \bmod p^\alpha.$$

For the special case when p is 2 this result can be extended. For this case (1) becomes

$$a = 1 + 2h.$$

Squaring we have

$$a^2 = 1 + 4h(h+1).$$

Hence,

$$a^2 = 1 + 8h_1, \tag{3}$$

where h_1 is an integer. Therefore

$$a^2 \equiv 1 \mod 2^3.$$

Squaring (3) we have

$$a^{2^2} = 1 + 2^4 h_2;$$

or

$$a^{2^2} \equiv 1 \mod 2^4.$$

It is now easy to see that we shall have in general

$$a^{2^{\alpha-2}} \equiv 1 \mod 2^\alpha$$

if $\alpha > 2$. That is,

$$a^{\frac{1}{2}\phi(2^\alpha)} \equiv 1 \mod 2^\alpha \text{ if } \alpha > 2.$$

Now in terms of the ϕ-function let us define a new function $\lambda(m)$ as follows:

$$\lambda(2^\alpha) = \phi(2^\alpha) \quad \text{if} \quad \alpha = 0, 1, 2;$$

$$\lambda(2^\alpha) = \tfrac{1}{2}\phi(2^\alpha) \quad \text{if} \quad \alpha > 2;$$

$$\lambda(p^\alpha) = \phi(p^\alpha) \quad \text{if } p \text{ is an odd prime;}$$

$$\lambda(2^\alpha p_1^{\alpha_1} p_2^{\alpha_2} \ldots p_n^{\alpha_n}) \quad = M,$$

where M is the least common multiple of

$$\lambda(2^\alpha), \quad \lambda(p_1^{\alpha_1}), \quad \lambda(p_2^{\alpha_2}), \ldots, \quad \lambda(p_n^{\alpha_n}),$$

$2, p_1, p_2, \ldots, p_n$ being different primes.

Denote by m the number

$$m = 2^\alpha p_1^{\alpha_1} p_2^{\alpha_2} \ldots p_n^{\alpha_n}.$$

Let a be any number prime to m. From our preceding results we have

$$a^{\lambda(2^\alpha)} \equiv 1 \mod 2^\alpha,$$

$$a^{\lambda(p_1^{\alpha_1})} \equiv 1 \mod p_1^{\alpha_1},$$

$$a^{\lambda(p_2^{\alpha_2})} \equiv 1 \mod p_2^{\alpha_2},$$

$$\cdot \quad \cdot \quad \cdot \quad \cdot \quad \cdot \quad \cdot \quad \cdot \quad \cdot$$

$$a^{\lambda(p_n^{\alpha_n})} \equiv 1 \mod p_n^{\alpha_n}.$$

Now any one of these congruences remains true if both of its members are raised to the same positive integral power, whatever that power may be. Then let us raise both members of the first congruence to the power $\lambda(m)/\lambda(2^\alpha)$; both members of the second congruence to the power $\lambda(m)/\lambda(p_1^{\alpha_1})$; . . . ; both members of the last congruence to the power $\lambda(m)/\lambda(p_n^{\alpha_n})$. Then we have

$$a^{\lambda(m)} \equiv 1 \bmod 2^\alpha,$$

$$a^{\lambda(m)} \equiv 1 \bmod p_1^{\alpha_1},$$

$$\cdot \quad \cdot \quad \cdot \quad \cdot \quad \cdot \quad \cdot$$

$$a^{\lambda(m)} \equiv 1 \bmod p_n^{\alpha_n}.$$

From these congruences we have immediately

$$a^{\lambda(m)} \equiv 1 \bmod m.$$

We may state this result in full in the following theorem:

If a and m are any two relatively prime positive integers, the congruence

$$a^{\lambda(m)} \equiv 1 \bmod m$$

is satisfied.

As an excellent example to show the possible difference between the exponent $\lambda(m)$ in this theorem and the exponent $\phi(m)$ in Fermat's general theorem, let us take

$$m = 2^6 \cdot 3^3 \cdot 5 \cdot 7 \cdot 13 \cdot 17 \cdot 19 \cdot 37 \cdot 73.$$

Here

$$\lambda(m) = 2^4 \cdot 3^2, \quad \phi(m) = 2^{21} \cdot 3^{10}.$$

In a later chapter we shall show that there is no exponent ν less than $\lambda(m)$ for which the congruence

$$a^\nu \equiv 1 \bmod m$$

is verified for every integer a prime to m.

From our theorem, as stated above, Fermat's general theorem follows as a corollary, since $\lambda(m)$ is obviously a factor of $\phi(m)$,

$$\phi(m) = \phi(2^\alpha)\phi(p_1^{\alpha_1}) \ . \ . \ . \ \phi(p_n^{\alpha_n}).$$

EXERCISES

1. Show that $a^{16} \equiv 1 \bmod 16320$, for every a which is prime to 16320.

2. Show that $a^{12} \equiv 1 \bmod 65520$, for every a which is prime to 65520.

3*. Find one or more composite numbers P such that

$$a^{P-1} \equiv 1 \bmod P$$

for every a prime to P. (Compare this problem with the next section.)

§ 29. On the Converse of Fermat's Simple Theorem

The fact that the converse of Wilson's theorem is a true proposition leads one naturally to inquire whether the converse of Fermat's simple theorem is true. Thus, we may ask the question: Does the existence of the congruence $2^{n-1} \equiv 1 \bmod n$ require that n be a prime number? The Chinese answered this question in the affirmative and the answer passed unchallenged among them for many years. An example is sufficient to show that the theorem is not true. We shall show that

$$2^{340} \equiv 1 \bmod 341$$

although $341, = 11 \cdot 31$, is not a prime number. Now $2^{10} - 1 = 3 \cdot 11 \cdot 31$. Hence $2^{10} \equiv 1 \bmod 341$. Hence $2^{340} \equiv 1 \bmod 341$. From this it follows that the direct converse of Fermat's theorem is not true. The following theorem, however, which is a converse with an extended hypothesis, is readily proved.

If there exists an integer a such that

$$a^{n-1} \equiv 1 \bmod n$$

and if further there does not exist an integer ν less than $n-1$ such that

$$a^{\nu} \equiv 1 \bmod n,$$

then the integer n is a prime number.

For, if n is not prime, $\phi(n) < n-1$. Then for $\nu = \phi(n)$ we have $a^{\nu} \equiv 1 \bmod n$, contrary to the hypothesis of the theorem.

§ 30. Application of Previous Results to Linear Congruences

The theorems of the present chapter afford us a ready means of writing down a solution of the congruence

$$ax \equiv c \bmod m.$$

We shall consider only the case in which a and m are relatively prime, since the general case is easily reducible to this one, as we saw in the preceding chapter.

Since a and m are relatively prime we have the congruences

$$a^{\lambda(m)} \equiv 1, \quad a^{\phi(m)} \equiv 1 \bmod m.$$

Hence either of the numbers x,

$$x = ca^{\lambda(m)-1}, \quad x = ca^{\phi(m)-1},$$

is a representative of the solution of (1). Hence the following theorem:

If $$ax \equiv c \bmod m$$

is any linear congruence in which a and m are relatively prime, then either of the numbers x,

$$x = ca^{\lambda(m)-1}, \quad x = ca^{\phi(m)-1},$$

is a representative of the solution of the congruence.

The former representative of the solution is the more convenient of the two, since the power of a is in general much less in this case than in the other.

EXERCISE

Find a solution of $7x \equiv 1 \bmod 2^6 \cdot 3 \cdot 5 \cdot 17$. Note the greater facility in applying the first of the above representatives of the solution rather than the second.

§ 31. Application of the Preceding Results to the Theory of Quadratic Residues

In this section we shall apply the preceding results of this chapter to the problem of finding the solutions of congruences of the form

$$\alpha z^2+\beta z+\gamma\equiv 0 \bmod \mu$$

where α, β, γ, μ are integers. These are called quadratic congruences.

The problem of the solution of the quadratic congruence (1) can be reduced to that of the solution of a simpler form of congruence as follows: Congruence (1) is evidently equivalent to the congruence

$$4\alpha^2z^2+4\alpha\beta z+4\alpha\gamma\equiv 0 \bmod 4\alpha\mu. \tag{1}$$

But this may be written in the form

$$(2\alpha z+\beta)^2\equiv\beta^2-4\alpha\gamma \bmod 4\alpha\mu.$$

Now if we put

$$2\alpha z+\beta\equiv x \bmod 4\alpha\mu \tag{2}$$

and

$$\beta^2-4\alpha\gamma=a,\quad 4\alpha\mu=m,$$

we have

$$x^2\equiv a \bmod m. \tag{3}$$

We have thus reduced the problem of solving the general congruence (1) to that of solving the binomial congruence (3) and the linear congruence (2). The solution of the latter may be effected by means of the results of § 30. We shall therefore confine ourselves now to a study of congruence (3). We shall make a further limitation by assuming that a and m are relatively prime, since it is obvious that the more general case is readily reducible to this one.

The example

$$x^2\equiv 3 \bmod 5$$

shows at once that the congruence (3) does not always have a solution. First of all, then, it is necessary to find out in what cases (3) has a solution. Before taking up the question it will be convenient to introduce some definitions.

Definitions. An integer a is said to be a quadratic residue modulo m or a quadratic non-residue modulo m according as the congruence

$$x^2 \equiv a \bmod m$$

has or has not a solution. We shall confine our attention to the case when $m > 2$.

We shall now prove the following theorem:

I. *If a and m are relatively prime integers, a necessary condition that a is a quadratic residue modulo m is that*

$$a^{\frac{1}{2}\lambda(m)} \equiv 1 \bmod m.$$

Suppose that the congruence $x^2 \equiv a \bmod m$ has the solution $x = \alpha$. Then $\alpha^2 \equiv a \bmod m$. Hence

$$\alpha^{\lambda(m)} \equiv a^{\frac{1}{2}\lambda(m)} \bmod m.$$

Since a is prime to m it is clear from $\alpha^2 \equiv a \bmod m$ that α is prime to m. Hence $\alpha^{\lambda(m)} \equiv 1 \bmod m$. Therefore we have

$$1 \equiv a^{\frac{1}{2}\lambda(m)} \bmod m.$$

That is, this is a necessary condition in order that a shall be a quadratic residue modulo m.

In a similar way one may prove the following theorem:

II. *If a and m are relatively prime integers, a necessary condition that a is a quadratic residue modulo m is that*

$$a^{\frac{1}{2}\phi(m)} \equiv 1 \bmod m.$$

When m is a prime number p each of the above results takes the following form: If a is prime to p and is a quadratic residue modulo p, then

$$a^{\frac{1}{2}(p-1)} \equiv 1 \bmod p.$$

We shall now prove the following more complete theorem, without the use of I or II.

III. *If p is an odd prime number and a is an integer not divisible by p, then a is a quadratic residue or a quadratic non-residue modulo p according as*

$$a^{\frac{1}{2}(p-1)} \equiv +1 \quad \text{or} \quad a^{\frac{1}{2}(p-1)} \equiv -1 \bmod p.$$

This is called Euler's criterion.

Given a number a, not divisible by p, we have to determine whether or not the congruence

$$x^2 \equiv a \bmod p$$

has a solution. Let r be any number of the set

$$1, 2, 3, \ldots, p-1 \qquad (A)$$

and consider the congruence

$$rx \equiv a \bmod p.$$

This has always one and just one solution x equal to a number s of the set (A). Two cases can arise: either for every r of the set (A) the corresponding s is different from r or for some r of the set (A) the corresponding s is equal to r. The former is the case when a is a quadratic non-residue modulo p; the latter is the case when a is a quadratic residue modulo p. We consider the two cases separately.

In the first case the numbers of the set (A) go in pairs such that the product of the numbers in the pair is congruent to a modulo p. Hence, taking the product of all $\frac{1}{2}(p-1)$ pairs, we have

$$1 \cdot 2 \cdot 3 \ldots \overline{p-1} \equiv +a^{\frac{1}{2}(p-1)} \bmod p.$$

But

$$1 \cdot 2 \cdot 3 \ldots \overline{p-1} \equiv -1 \bmod p.$$

Hence

$$a^{\frac{1}{2}(p-1)} \equiv -1 \bmod p,$$

whence the truth of one part of the theorem.

In the other case, namely that in which some r and corresponding s are equal, we have for this r

$$r^2 \equiv a \bmod p$$

and

$$(p-r)^2 \equiv a \bmod p.$$

Since $x^2 \equiv a \bmod p$ has at most two solutions it follows that all the integers in the set (A) except r and $p-r$ fall in pairs such that the product of the numbers in each pair is congruent to a modulo p. Hence, taking the product of all these pairs, which are $\frac{1}{2}(p-1)-1$ in number, and multiplying by $r(p-r)$ we have

$$\begin{aligned} 1\cdot 2\cdot 3\ \ldots\ \overline{p-1} &\equiv (p-r)ra^{\frac{1}{2}(p-1)-1} \bmod p \\ &\equiv -r^2 a^{\frac{1}{2}(p-1)-1} \bmod p \\ &\equiv -a\cdot a^{\frac{1}{2}(p-1)-1} \bmod p \\ &\equiv -a^{\frac{1}{2}(p-1)} \bmod p. \end{aligned}$$

Since $1\cdot 2\cdot 3\ \ldots\ \overline{p-1} \equiv -1 \bmod p$ we have

$$a^{\frac{1}{2}(p-1)} \equiv +1 \bmod p;$$

whence the truth of another part of the theorem.

Thus the proof of the entire theorem is complete.

CHAPTER V.

PRIMITIVE ROOTS MODULO m.

§ 32. Exponent of an Integer Modulo m

Let

$$a_1, a_2, \ . \ . \ . \ , a_{\phi(m)} \qquad (A)$$

be the set of $\phi(m)$ positive integers not greater than m and prime to m; and let a denote any integer of the set (A). Now any positive integral power of a is prime to m and hence is congruent modulo m to a number of the set (A). Hence, among all the powers of a there must be two, say a^n and a^ν, $n>\nu$, which are congruent to the same integer of the set (A). These two powers are then congruent to each other; that is,

$$a^n \equiv a^\nu \bmod m.$$

Since a^ν is prime to m the members of this congruence may be divided by a^ν. Thus we have

$$a^{n-\nu} \equiv 1 \bmod m.$$

That is, among the powers of a there is one at least which is congruent to 1 modulo m.

Now, in the set of all powers of a which are congruent to 1 modulo m there is one in which the exponent is less than in any other of the set. Let the exponent of this power be d, so that a^d is the lowest power of a such that

$$a^d \equiv 1 \bmod m.$$

We shall now show that if $a^\alpha \equiv 1 \bmod m$, then α is a multiple of d. Let us write

$$\alpha = d\delta + \beta, \quad 0 \leq \beta < d.$$

Then

$$a^\alpha \equiv 1 \bmod m, \tag{2}$$

$$a^{d\delta} \equiv 1 \bmod m, \tag{3}$$

the last congruence being obtained by raising (1) to the power δ. From (3) we have

$$a^{d\delta+\beta} \equiv a^\beta \bmod m;$$

or

$$a^\beta \equiv 1 \bmod m.$$

Hence $\beta = 0$, for otherwise d is not the exponent of the lowest power of a which is congruent to 1 modulo m. Hence d is a divisor of α.

These results may be stated as follows:

I. *If m is any integer and a is any integer prime to m, then there exists an integer d such that*

$$a^d \equiv 1 \bmod m$$

while there is no integer β less than d for which

$$a^\beta \equiv 1 \bmod m.$$

Further, a necessary and sufficient condition that

$$a^\nu \equiv 1 \bmod m$$

is that ν is a multiple of d.

Definition. The integer d which is thus uniquely determined when the two relatively prime integers a and m are given is called the exponent of a modulo m. Also, d is said to be the exponent to which a belongs modulo m.

Now, in every case we have

$$a^{\phi(m)} \equiv 1, \quad a^{\lambda(m)} \equiv 1 \bmod m,$$

if a and m are relatively prime. Hence from the preceding theorem we have at once the following:

II. *The exponent d to which a belongs modulo m is a divisor of both $\phi(m)$ and $\lambda(m)$.*

§ 33. Another Proof of Fermat's General Theorem

In this section we shall give an independent proof of the theorem that the exponent d of a modulo m is a divisor of $\phi(m)$; from this result we have obviously a new proof of Fermat's theorem itself.

We retain the notation of the preceding section. We shall first prove the following theorem:

I. *The numbers*

$$1, a, a^2, \ldots, a^{d-1} \qquad (A)$$

are incongruent each to each modulo m.

For, if $a^\alpha \equiv a^\beta \bmod m$, where $0 \leqq \alpha < d$ and $0 \leqq \beta < d$, $\alpha > \beta$, we have $a^{\alpha-\beta} \equiv 1 \bmod m$, so that d is not the exponent to which a belongs modulo m, contrary to hypothesis.

Now any number of the set (A) is congruent to some number of the set

$$a_1, a_2, \ldots, a_{\phi(m)}. \qquad (B)$$

Let us undertake to separate the numbers (B) into classes after the following manner: Let the first class consist of the numbers

$$\text{(I)} \qquad \alpha_0, \alpha_1, \alpha_2, \ldots, \alpha_{d-1},$$

where α_i is the number of the set (B) to which a^i is congruent modulo m.

If the class (I) does not contain all the numbers of the set (B), let a_i be any number of the set (B) not contained in (I) and form the following set of numbers:

$$\text{(II}'\text{)} \qquad \alpha_0 a_i, \alpha_1 a_i, \alpha_2 a_i, \ldots, \alpha_{d-1} a_i.$$

We shall now show that no number of this set is congruent to a number of class (I). For, if so, we should have a congruence of the form

$$a_i\alpha_j \equiv \alpha_k \bmod m;$$

hence

$$a_i a^j \equiv a^k \bmod m,$$

so that

$$a_i a^d \equiv a^{k+d-j} \bmod m;$$

or

$$a_i \equiv a^{k+d-j} \bmod m,$$

so that a_i would belong to the set (I) contrary to hypothesis.

Now the numbers of the set (II′) are all congruent to numbers of the set (B); and no two are congruent to the same number of this set. For, if so, we should have two numbers of (II′) congruent; that is, $\alpha_k a_i \equiv \alpha_j a_i \bmod m$, or $\alpha_k \equiv \alpha_j \bmod m$; and this we have seen to be impossible.

Now let the numbers of the set (B) to which the numbers of the set (II′) are congruent be in order the following:

(II) $$\beta_0, \beta_1, \beta_2, \ldots, \beta_{d-1}.$$

These numbers constitute our class (II).

If classes (I) and (II) do not contain all the numbers of the set (B), let a_j be a number of the set (B) not contained in either of the classes (I) and (II): and form the set of numbers

(III′) $$\alpha_0 a_j, \alpha_1 a_j, \alpha_2 a_j, \ldots, \alpha_{d-1} a_j.$$

Just as in the preceding case it may be shown that no number of this set is congruent to a number of class (I) and that the numbers of (III′) are incongruent each to each. We shall also show that no number of (III′) is congruent to a number of class (II). For, if so, we should have $a_k a_j \equiv \beta_l \bmod m$. Hence $a^k a_j \equiv a^l a_i \bmod m$; or $a_j \equiv a^{l+d-k} a_i \bmod m$, from which it follows that a_j is of class (II), contrary to hypothesis.

Now let the numbers of the set (B) to which the numbers of the set (III′) are congruent be in order the following:

(III) $$\gamma_0, \gamma_1, \gamma_2, \ldots, \gamma_{d-1}.$$

These numbers form our class (III).

It is now evident that the process may be continued until all the numbers of the set (B) have been separated into classes, each class containing d integers, thus:

$$
\begin{array}{ll}
\text{(I)} & \alpha_0,\ \alpha_1,\ \alpha_2,\ \ldots,\ \alpha_{d-1},\\
\text{(II)} & \beta_0,\ \beta_1,\ \beta_2,\ \ldots,\ \beta_{d-1},\\
\text{(III)} & \gamma_0,\ \gamma_1,\ \gamma_2,\ \ldots,\ \gamma_{d-1},\\
 & \ldots\ldots\ldots\ldots\\
(\quad) & \lambda_0,\ \lambda_1,\ \lambda_2,\ \ldots,\ \lambda_{d-1}.
\end{array}
$$

The set (B), which consists of $\phi(m)$ integers, has thus been separated into classes, each class containing d integers. Hence we conclude that d is a divisor of $\phi(m)$. Thus we have a second proof of the theorem:

II. *If a and m are any two relatively prime integers and d is the exponent to which a belongs modulo m, then d is a divisor of $\phi(m)$.*

In our classification of the numbers (B) into the rectangular array above we have proved much more than theorem II; in fact, theorem II is to be regarded as one only of the consequences of the more general result contained in the array.

If we raise each member of the congruence

$$a^d \equiv 1 \bmod m$$

to the (integral) power $\phi(m)/d$, the preceding theorem leads immediately to an independent proof of Fermat's general theorem.

§ 34. Definition of Primitive Roots

Definition. Let a and m be two relatively prime integers. If the exponent to which a belongs modulo m is $\phi(m)$, a is said to be a primitive root modulo m (or a primitive root of m).

In a previous chapter we saw that the congruence

$$a^{\lambda(m)} \equiv 1 \bmod m$$

is verified by every pair of relatively prime integers a and m. Hence, primitive roots can exist only for such a modulus m as satisfies the equation

$$\phi(m) = \lambda(m). \qquad (1)$$

We shall show later that this is also sufficient for the existence of primitive roots.

From the relation which exists in general between the ϕ-function and the λ-function in virtue of the definition of the latter, it follows that (1) can be satisfied only when m is a prime power or is twice an odd prime power.

Suppose first that m is a power of 2, say $m = 2^\alpha$. Then (1) is satisfied only if $\alpha = 0, 1, 2$. For $\alpha = 0$ or 1, 1 itself is a primitive root. For $\alpha = 2$, 3 is a primitive root. We have therefore left to examine only the cases

$$m = p^\alpha, \quad m = 2p^\alpha$$

where p is an odd prime number. The detailed study of these cases follows in the next sections.

§ 35. Primitive Roots Modulo p.

We have seen that if p is a prime number and d is the exponent to which a belongs modulo p, then d is a divisor of $\phi(p) = p - 1$. Now, let

$$d_1, d_2, d_3, \ldots, d_r$$

be all the divisors of $p - 1$ and let $\psi(d_i)$ denote the number of integers of the set

$$1, 2, 3, \ldots, p-1$$

which belong to the exponent d_i. If there is no integer of the set belonging to this exponent, then $\psi(d_i) = 0$.

Evidently every integer of the set belongs to some one and only one of the exponents $d_1, d_2, \ldots, d_r$. Hence we have the relation

$$\psi(d_1)+\psi(d_2)+\ldots+\psi(d_r)=p-1. \qquad (1)$$

But

$$\phi(d_1)+\phi(d_2)+\ldots+\phi(d_r)=p-1. \qquad (2)$$

If then we can show that

$$\psi(d_i) \leqq \phi(d_i) \qquad (3)$$

for $i=1, 2, \ldots, r$, it will follow from a comparison of (1) and (2) that

$$\psi(d_i)=\phi(d_i).$$

Accordingly, we shall examine into the truth of (3).

Now the congruence

$$x^{d_i}\equiv 1 \bmod p \qquad (4)$$

has not more than d_i roots. If no root of this congruence belongs to the exponent d_i, then $\psi(d_i)=0$ and therefore in this case we have $\psi(d_i)<\phi(d_i)$. On the other hand if a is a root of (4) belonging to the exponent d_i, then

$$a, a^2, a^3, \ldots, a^{d_i} \qquad (5)$$

are a set of d_i incongruent roots of (4); and hence they are the complete set of roots of (4).

But it is easy to see that a^k does or does not belong to the exponent d_i according as k is or is not prime to d_i; for, if a^k belongs to the exponent t, then t is the least integer such that kt is a multiple of d_i. Consequently the number of roots in the set (5) belonging to the exponent d_i is $\phi(d_i)$. That is, in this case $\psi(d_i)=\phi(d_i)$. Hence in general $\psi(d_i)\leqq\phi(d_i)$. Therefore from (1) and (2) we conclude that

$$\psi(d_i)=\phi(d_i), \qquad i=1, 2, \ldots, r.$$

The result thus obtained may be stated in the form of the following theorem:

I. *If p is a prime number and d is any divisor of $p-1$, then the number of integers belonging to the exponent d modulo p is $\phi(d)$.*

In particular:

II. *There exist primitive roots modulo p and their number is $\phi(p-1)$.*

§ 36. Primitive Roots Modulo p^α, p an Odd Prime

In proving that there exist primitive roots modulo p^α, where p is an odd prime and $\alpha > 1$, we shall need the following theorem:

I. *There always exists a primitive root γ modulo p for which $\gamma^{p-1}-1$ is not divisible by p^2.*

Let g be any primitive root modulo p. If $g^{p-1}-1$ is not divisible by p^2 our theorem is verified. Then suppose that $g^{p-1}-1$ is divisible by p^2, so that we have

$$g^{p-1}-1=kp^2$$

where k is an integer. Then put

$$\gamma=g+xp$$

where x is an integer. Then $\gamma\equiv g \bmod p$, and hence

$$\gamma^h \equiv g^h \bmod p;$$

whence we conclude that γ is a primitive root modulo p. But

$$\gamma^{p-1}-1=g^{p-1}-1+\frac{p-1}{1!}g^{p-2}xp+\frac{(p-1)(p-2)}{2!}g^{p-3}x^2p^2+\ \ldots$$

$$=p\left(kp+\frac{p-1}{1!}g^{p-2}x+\frac{(p-1)(p-2)}{2!}g^{p-3}x^2p+\ \ldots\right).$$

Hence

$$\gamma^{p-1}-1\equiv p(-g^{p-2}x) \bmod p^2.$$

Therefore it is evident that x can be so chosen that $\gamma^{p-1}-1$ is not divisible by p^2. Hence there exists a primitive root γ modulo p such that $\gamma^{p-1}-1$ is not divisible by p^2. Q. E. D.

We shall now prove that this integer γ is a primitive root modulo p^α, where α is any positive integer.

If

$$\gamma^k \equiv 1 \bmod p,$$

then k is a multiple of $p-1$, since γ is a primitive root modulo p. Hence, if

$$\gamma^k \equiv 1 \bmod p^\alpha,$$

then k is a multiple of $p-1$.

Now, write

$$\gamma^{p-1} = 1 + hp.$$

Since $\gamma^{p-1}-1$ is not divisible by p^2, it follows that h is prime to p. If we raise each member of this equation to the power $\beta p^{\alpha-2}$, $\alpha \geqq 2$, we have

$$\gamma^{\beta p^{\alpha-2}(p-1)} = 1 + \beta p^{\alpha-1}h + p^\alpha I,$$

where I is an integer. Then if

$$\gamma^{\beta p^{\alpha-2}(p-1)} \equiv 1 \bmod p^\alpha,$$

β must be divisible by p. Therefore the exponent of the lowest power of γ which is congruent to 1 modulo p^α is divisible by $p^{\alpha-1}$. But we have seen that this exponent is also divisible by $p-1$. Hence the exponent of γ modulo p^α is $p^{\alpha-1}(p-1)$, since $\phi(p^\alpha) = p^{\alpha-1}(p-1)$. That is, γ is a primitive root modulo p^α.

It is easy to see that no two numbers of the set

$$\gamma, \gamma^2, \gamma^3, \ldots, \gamma^{p^{\alpha-1}(p-1)} \qquad (A)$$

are congruent modulo p^α; for, if so, γ would belong modulo p^α to an exponent less than $p^{\alpha-1}(p-1)$ and would therefore not be a primitive root modulo p^α. Now every number in the set

(A) is prime to p^α; their number is $\phi(p^\alpha)=p^{\alpha-1}(p-1)$. Hence the numbers of the set (A) are congruent in some order to the numbers of the set (B):

$$a_1, a_2, a_3, \ldots, a_{p^{\alpha-1}(p-1)}, \qquad (B)$$

where the integers (B) are the positive integers less than p^α and prime to p^α.

But any number of the set (B) is a solution of the congruence

$$x^{p^{\alpha-1}(p-1)} \equiv 1 \bmod p^\alpha. \qquad (1)$$

Further, every solution of this congruence is prime to p^α. Hence the integers (B) are a complete set of solutions of (1). Therefore the integers (A) are a complete set of solutions of (1). But it is easy to see that an integer γ^k of the set (A) is or is not a primitive root modulo p^α according as k is or is not prime to $p^{\alpha-1}(p-1)$. Hence the number of primitive roots modulo p^α is $\phi\{p^{\alpha-1}(p-1)\}$.

The results thus obtained may be stated as follows:

II. *If p is any odd prime number and α is any positive integer, then there exist primitive roots modulo p^α and their number is $\phi\{\phi(p^\alpha)\}$.*

§ 37. Primitive Roots Modulo $2p^\alpha$, p an Odd Prime

In this section we shall prove the following theorem:

If p is any odd prime number and α is any positive integer, then there exist primitive roots modulo $2p^\alpha$ and their number is $\phi\{\phi(2p^\alpha)\}$.

Since $2p^\alpha$ is even it follows that every primitive root modulo $2p^\alpha$ is an odd number. Any odd primitive root modulo p^α is obviously a primitive root modulo $2p^\alpha$. Again, if γ is an even primitive root modulo p^α then $\gamma+p^\alpha$ is a primitive root modulo $2p^\alpha$. It is evident that these two classes contain (without repetition) all the primitive roots modulo $2p^\alpha$. Hence the theorem follows as stated above.

§ 38. Recapitulation

The results which we have obtained in §§ 34–37 inclusive may be gathered into the following theorem:

In order that there shall exist primitive roots modulo m, it is necessary and sufficient that m shall have one of the values

$$m = 1,\ 2,\ 4,\ p^\alpha,\ 2p^\alpha$$

where p is an odd prime and α is a positive integer.

If m has one of these values then the number of primitive roots modulo m is $\phi\{\phi(m)\}$.

§ 39. Primitive λ-roots

In the preceding sections of this chapter we have developed the theory of primitive roots in the way in which it is usually presented. But if one approaches the subject from a more general point of view the results which may be obtained are more general and at the same time more elegant. It is our purpose in this section to develop the more general theory.

We have seen that if a and m are any two relatively prime positive integers, then

$$a^{\lambda(m)} \equiv 1 \bmod m.$$

Consequently there is no integer belonging modulo m to an exponent greater than $\lambda(m)$. It is natural to enquire if there are any integers a which belong to the exponent $\lambda(m)$. It turns out that the question is to be answered in the affirmative, as we shall show. Accordingly, we introduce the following definition:

Definition. If $a^{\lambda(m)}$ is the lowest power of a which is congruent to 1 modulo m, a is said to be a primitive λ-root modulo m. We shall also say that it is a primitive λ-root of the congruence $x^{\lambda(m)} \equiv 1 \bmod m$. To distinguish we may speak of the usual primitive root as a primitive ϕ-root modulo m.

From the theory of primitive ϕ-roots already developed it follows that primitive λ-roots always exist when m is a power of any odd prime, and also when $m = 1, 2, 4$; for, for such values of m we have $\lambda(m) = \phi(m)$.

We shall next show that primitive λ-roots exist when $m = 2^\alpha$, $\alpha > 2$, by showing that 5 is such a root. It is necessary and sufficient to prove that 5 belongs modulo 2^α to the exponent $2^{\alpha-2} = \lambda(2^\alpha)$. Let d be the exponent to which 5 belongs modulo 2^α. Then from theorem II of § 32 it follows that d is a divisor of $2^{\alpha-2} = \lambda(2^\alpha)$. Hence if d is different from $2^{\alpha-2}$ it is $2^{\alpha-3}$ or is a divisor of $2^{\alpha-3}$. Hence if we can show that $5^{2^{\alpha-3}}$ is not congruent to 1 modulo 2^α we will have proved that 5 belongs to the exponent $2^{\alpha-2}$. But, clearly,

$$5^{2^{\alpha-3}} = (1 + 2^2)^{2^{\alpha-3}} = 1 + 2^{\alpha-1} + I \cdot 2^\alpha,$$

where I is an integer. Hence

$$5^{2^{\alpha-3}} \not\equiv 1 \bmod 2^\alpha.$$

Hence 5 belongs modulo 2^α to the exponent $\lambda(2^\alpha)$.

By means of these special results we are now in position to prove readily the following general theorem which includes them as special cases:

I. *For every congruence of the form*

$$x^{\lambda(m)} \equiv 1 \bmod m$$

a solution g exists which is a primitive λ-root, and for any such solution g there are $\phi\{\lambda(m)\}$ primitive roots congruent to powers of g.

If any primitive λ-root g exists, g^ν is or is not a primitive λ-root according as ν is or is not prime to $\lambda(m)$; and therefore the number of primitive λ-roots which are congruent to powers of any such root g is $\phi\{\lambda(m)\}$.

The existence of a primitive λ-root in every case may easily be shown by induction. In case m is a power of a prime the theorem has already been established. We will suppose that it is true when m is the product of powers of r different primes and show that it is true when m is the product of powers of $r+1$ different primes; from this will follow the theorem in general.

Put $m=p_1^{\alpha_1}p_2^{\alpha_2}\ldots p_r^{\alpha_r}p_{r+1}^{\alpha_{r+1}}$, $n=p_1^{\alpha_1}p_2^{\alpha_2}\ldots p_r^{\alpha_r}$,

and let h be a primitive λ-root of

$$x^{\lambda(n)}\equiv 1 \bmod n. \tag{1}$$

Then

$$h+ny$$

is a form of the same root if y is an integer.

Likewise, if c is any primitive λ-root of

$$x^{\lambda(p_{r+1}^{\alpha_{r+1}})}\equiv 1 \bmod p_{r+1}^{\alpha_{r+1}} \tag{2}$$

a form of this root is

$$c+p_{r+1}^{\alpha_{r+1}}z$$

where z is any integer.

Now, if y and z can be chosen so that

$$h+ny=c+p_{r+1}^{\alpha_{r+1}}z$$

the number in either member of this equation will be a common primitive λ-root of congruences (1) and (2); that is, a common primitive λ-root of the two congruences may always be obtained provided that the equation

$$p_1^{\alpha_1}\ldots p_r^{\alpha_r}y-p_{r+1}^{\alpha_{r+1}}z=c-h$$

has always a solution in which y and z are integers. That this equation has such a solution follows readily from theorem III of § 9; for, if $c-h$ is replaced by 1, the new equation has a solution $\bar{y}$, $\bar{z}$; and therefore for y and z we may take $y=\bar{y}(c-h)$, $z=\bar{z}(c-h)$.

Now let g be a common primitive λ-root of congruences (1) and (2) and write

$$g^{\nu}\equiv 1 \bmod m,$$

where ν is to be the smallest exponent for which the congruence is true. Since g is a primitive λ-root of (1) ν is a multiple of $\lambda(p_1^{\alpha_1} \ldots p_r^{\alpha_r})$. Since g is a primitive λ-root of (2) ν is a multiple of $\lambda\left(p_{r+1}^{\alpha_{r+1}}\right)$. Hence it is a multiple of $\lambda(m)$. But $g^{\lambda(m)} \equiv 1 \bmod m$; therefore $\nu = \lambda(m)$. That is, g is a primitive λ-root modulo m.

The theorem as stated now follows at once by induction.

There is nothing in the preceding argument to indicate that the primitive λ-roots modulo m are all in a single set obtained by taking powers of some root g; in fact it is not in general true when m contains more than one prime factor.

By taking powers of a primitive λ-root g modulo m one obtains $\phi\{\lambda(m)\}$ different primitive λ-roots modulo m. It is evident that if γ is any one of these primitive λ-roots, then the same set is obtained again by taking the powers of γ. We may say then that the set thus obtained is the set belonging to g.

II. *If $\lambda(m) > 2$ the product of the $\phi\{\lambda(m)\}$ primitive λ-roots in the set belonging to any primitive λ-root g is congruent to 1 modulo m.*

These primitive λ-roots are

$$g, g^{c_1}, g^{c_2}, \ldots, g^{c_\mu}$$

where

$$1, c_1, c_2, \ldots, c_\mu$$

are the integers less than $\lambda(m)$ and prime to $\lambda(m)$. If any one of these is c another is $\lambda(m) - c$, since $\lambda(m) > 2$. Hence

$$1 + c_1 + c_2 + \ldots + c_\mu \equiv 0 \bmod \lambda(m).$$

Therefore

$$g^{1 + c_1 + c_2 + \ldots + c_\mu} \equiv 1 \bmod m.$$

From this the theorem follows.

COROLLARY. *The product of all the primitive λ-roots modulo m is congruent to 1 modulo m when $\lambda(m) > 2$.*

EXERCISES

1. If x_1 is the largest value of x satisfying the equation $\lambda(x)=a$, where a is a given integer, then any solution x_2 of the equation is a factor of x_1.

2*. Obtain an effective rule for solving the equation $\lambda(x)=a$.

3*. Obtain an effective rule for solving the equation $\phi(x)=a$.

4. A necessary and sufficient condition that $a^{P-1}\equiv 1 \bmod P$ for every integer a prime to P is that $P\equiv 1 \bmod \lambda(P)$.

5. If $a^{P-1}\equiv 1 \bmod P$ for every a prime to P, then (1) P does not contain a square factor other than 1, (2) P either is prime or contains at least three different prime factors.

6. Let p be a prime number. If a is a root of the congruence $x^d\equiv 1 \bmod p$ and α is a root of the congruence $x^\delta\equiv 1 \bmod p$, then $a\alpha$ is a root of the congruence $x^{d\delta}\equiv 1 \bmod p$. If a is a primitive root of the first congruence and α of the second and if d and δ are relatively prime, then $a\alpha$ is a primitive root of the congruence $x^{d\delta}\equiv 1 \bmod p$.

CHAPTER VI

OTHER TOPICS

§ 40. Introduction

The theory of numbers is a vast discipline and no single volume can adequately treat of it in all of its phases. A short book can serve only as an introduction; but where the field is so vast such an introduction is much needed. That is the end which the present volume is intended to serve; and it will best accomplish this end if, in addition to the detailed theory already developed, some account is given of the various directions in which the matter might be carried further.

To do even this properly it is necessary to limit the number of subjects considered. Consequently we shall at once lay aside many topics of interest which would find a place in an exhaustive treatise. We shall say nothing, for instance, about the vast domain of algebraic numbers, even though this is one of the most fascinating subjects in the whole field of mathematics. Consequently, we shall not refer to any of the extensive theory connected with the division of the circle into equal parts. Again, we shall leave unmentioned many topics connected with the theory of positive integers; such, for instance, is the frequency of prime numbers in the ordered system of integers—a subject which contains in itself an extensive and elegant theory.

In §§ 41–44 we shall speak briefly of each of the following topics: theory of quadratic residues, Galois imaginaries, arithmetic forms, analytical theory of numbers. Each of these alone would require a considerable volume for its proper development. All that we can do is to indicate the nature of the problem in each case and in some cases to give a few of the fundamental results.

In the remaining three sections we shall give a brief introduction to the theory of Diophantine equations, developing some of the more elementary properties of certain special cases. We shall carry this far enough to indicate the nature of the problem connected with the now famous Last Theorem of Fermat. The earlier sections of this chapter are not required as a preliminary to reading this latter part.

§ 41. Theory of Quadratic Residues

Let a and m be any two relatively prime integers. In § 31 we agreed to say that a is a quadratic residue modulo m or a quadratic non-residue modulo m according as the congruence

$$x^2 \equiv a \bmod m$$

has or has not a solution. We saw that if m is chosen equal to an odd prime number p, then a is a quadratic residue modulo p or a quadratic non-residue modulo p according as

$$a^{\frac{1}{2}(p-1)} \equiv 1 \quad \text{or} \quad a^{\frac{1}{2}(p-1)} \equiv -1 \bmod p.$$

This is known as Euler's criterion.

It is convenient to employ the Legendre symbol

$$\left(\frac{a}{p}\right)$$

to denote the quadratic character of a with respect to p. This symbol is to have the value $+1$ or the value -1 according as a is a quadratic residue modulo p or a quadratic non-residue modulo p. We shall now derive some of the fundamental properties of this symbol, understanding always that the numbers in the numerator and the denominator are relatively prime.

From the definition of quadratic residues and non-residues it is obvious that

$$\left(\frac{a}{p}\right) = \left(\frac{b}{p}\right) \text{ if } a \equiv b \bmod p. \qquad (1)$$

It is easy to prove in general that

$$\left(\frac{a}{p}\right)\left(\frac{b}{p}\right)=\left(\frac{ab}{p}\right). \qquad (2)$$

This comes readily from Euler's criterion. We have to consider the three cases

$$\left(\frac{a}{p}\right)=+1,\quad \left(\frac{b}{p}\right)=+1;\quad \left(\frac{a}{p}\right)=+1,\quad \left(\frac{b}{p}\right)=-1;$$
$$\left(\frac{a}{p}\right)=-1,\quad \left(\frac{b}{p}\right)=-1.$$

The method will be sufficiently illustrated by the treatment of the last case. Here we have

$$a^{\frac{1}{2}(p-1)}\equiv -1 \bmod p,\quad b^{\frac{1}{2}(p-1)}\equiv -1 \bmod p.$$

Multiplying these two congruences together member by member we have

$$(ab)^{\frac{1}{2}(p-1)}\equiv 1 \bmod p,$$

whence

$$\left(\frac{ab}{p}\right)=1=\left(\frac{a}{p}\right)\left(\frac{b}{p}\right),$$

as was to be proved.

If m is any number prime to p and we write m as the product of factors

$$m=\epsilon\cdot 2^{\alpha}\cdot q'q''q''' \ \ldots$$

where q', q'', q''', ... are odd primes, α is zero or a positive integer and ϵ is $+1$ or -1 according as m is positive or negative, we have

$$\left(\frac{m}{p}\right)=\left(\frac{\epsilon}{p}\right)\left(\frac{2}{p}\right)^{\alpha}\left(\frac{q'}{p}\right)\left(\frac{q''}{p}\right)\left(\frac{q'''}{p}\right)\ \ldots, \qquad (3)$$

as one shows easily by repeated application of relation (2). Obviously,

$$\left(\frac{1}{p}\right)=1.$$

Hence, it follows from (3) that we can readily determine the quadratic character of m with respect to the odd prime p, that is, the value of

$$\left(\frac{m}{p}\right),$$

provided that we know the value of each of the expressions

$$\left(\frac{-1}{p}\right), \quad \left(\frac{2}{p}\right), \quad \left(\frac{q}{p}\right), \tag{4}$$

where q is an odd prime.

The first of these can be evaluated at once by means of Euler's criterion; for, we have

$$\left(\frac{-1}{p}\right) \equiv (-1)^{\frac{1}{2}(p-1)} \bmod p$$

and hence

$$\left(\frac{-1}{p}\right) = (-1)^{\frac{1}{2}(p-1)}.$$

Thus we have the following result: The number -1 is a quadratic residue of every prime number of the form $4k+1$ and a quadratic non-residue of every prime number of the form $4k+3$.

The value of the second symbol in (4) is given by the formula

$$\left(\frac{2}{p}\right) = (-1)^{\frac{1}{8}(p^2-1)}.$$

The theorem contained in this equation may be stated in the following words: The number 2 is a quadratic residue of every prime number of either of the forms $8k+1$, $8k+7$; it is a quadratic non-residue of every prime number of either of the forms $8k+3$, $8k+5$.

The proof of this result is not so immediate as that of the preceding one. To evaluate the third expression in (4) is still more difficult. We shall omit the demonstration in both of these cases. For the latter we have the very elegant relation

$$\left(\frac{p}{q}\right)\left(\frac{q}{p}\right) = (-1)^{\frac{1}{4}(p-1)(q-1)}.$$

This equation states the law which connects the quadratic character of q with respect to p with the quadratic character of p with respect to q. It is known as the Law of Quadratic Reciprocity. About fifty proofs of it have been given. Its history has been a very interesting one; see Bachmann's Niedere Zahlentheorie, Teil I, pp. 180–318, especially pp. 200–206.

For a further account of this beautiful and interesting subject we refer the reader to Bachmann, loc. cit., and to the memoirs to which this author gives reference.

§ 42. Galois Imaginaries

If one is working in the domain of real numbers the equation

$$x^2+1=0$$

has no solution; for there is no real number whose square is -1. If, however, one enlarges the "number system" so as to include not only all real numbers but all complex numbers as well, then it is true that every algebraic equation has a root. It is on account of the existence of this theorem for the enlarged domain that much of the general theory of algebra takes the elegant form in which we know it.

The question naturally arises as to whether we can make a similar extension in the case of congruences. The congruence

$$x^2 \equiv 3 \bmod 5$$

has no solution, if we employ the term solution in the sense in which we have so far used it. But we may if we choose introduce an imaginary quantity, or mark, j such that

$$j^2 \equiv 3 \bmod 5,$$

just as in connection with the equation $x^2+1=0$ we would introduce the symbol i having the property expressed by the equation

$$i^2=-1.$$

It is found to be possible to introduce in this way a general set of imaginaries satisfying congruences with prime moduli; and the new quantities or marks have the property of combining according to the laws of algebra.

The quantities so introduced are called Galois imaginaries.

We cannot go into a development of the important theory which is introduced in this way. We shall be content with indicating two directions in which it leads.

In the first place there is the general Galois field theory which is of fundamental importance in the study of certain finite groups. It may be developed from the point of view indicated here. An excellent exposition, along somewhat different lines, is to be found in Dickson's *Linear Groups with an Exposition of the Galois Field Theory*.

Again, the whole matter may be looked upon from the geometric point of view. In this way we are led to the general theory of finite geometries, that is, geometries in which there is only a finite number of points. For a development of the ideas which arise here see Veblen and Young's *Projective Geometry* and the memoir by Veblen and Bussey in the Transactions of the American Mathematical Society, vol. 7, pp. 241–259.

§ 43. Arithmetic Forms

The simplest arithmetic form is $ax+b$ where a and b are fixed integers different from zero and x is a variable integer. By varying x in this case we have the terms of an arithmetic progression. We have already referred to Dirichlet's celebrated theorem which asserts that the form $ax+b$ has an infinite number of prime values if only a and b are relatively prime. This is an illustration of one type of theorem connected with arithmetic forms in general, namely, those in which it is asserted that numbers of a given form have in addition a given property.

Another type of theorem is illustrated by a result stated in § 41, provided that we look at that result in the proper way. We saw that the number 2 is a quadratic residue of every prime of either of the forms $8k+1$ and $8k+7$ and a quadratic non-residue of every prime of either of the forms $8k+3$ and $8k+5$. We may state that result as follows: A given prime number of either of the forms $8k+1$ and $8k+7$ is a divisor of some number of the form x^2-2, where x is an integer; no prime number of either of the forms $8k+3$ and $8k+5$ is a divisor of a number of the form x^2-2, where x is an integer.

The result just stated is a theorem in a discipline of vast extent, namely, the theory of quadratic forms. Here a large number of questions arise among which are the following: What numbers can be represented in a given form? What is the character of the divisors of a given form? As a special case of the first we have the question as to what numbers can be represented as the sum of three squares. To this category belong also the following two theorems: Every positive integer is the sum of four squares of integers; every prime number of the form $4n+1$ may be represented (and in only one way) as the sum of two squares.

For an extended development of the theory of quadratic forms we refer the reader to Bachmann's Arithmetik der Quadratischen Formen of which the first part has appeared in a volume of nearly seven hundred pages.

It is clear that one may further extend the theory of arithmetic forms by investigating the properties of those of the third and higher degrees. Naturally the development of this subject has not been carried so far as that of quadratic forms; but there is a considerable number of memoirs devoted to various parts of this extensive field, and especially to the consideration of various special forms.

Probably the most interesting of these special forms are the following:

$$\alpha^n+\beta^n, \quad \frac{\alpha^n-\beta^n}{\alpha-\beta}=\alpha^{n-1}+\alpha^{n-2}\beta+ \ . \ . \ . \ +\beta^{n-1},$$

where α and β are relatively prime integers, or, more generally, where α and β are the roots of the quadratic equation $x^2-ux+v=0$ where u and v are relatively prime integers. A development of the theory of these forms has been given by the present author in a memoir published in 1913 in the Annals of Mathematics, vol. 13, pp. 30–70.

§ 44. Analytical Theory of Numbers

Let us consider the function

$$P(x) = \frac{1}{\prod_{k=0}^{\infty}(1-x^{2^k})}, \qquad |x| \leq \rho < 1.$$

It is clear that we have

$$P(x) = \prod_{k=0}^{\infty} \frac{1}{1-x^{2^k}} = \prod_{k=0}^{\infty} (1+x^{2^k}+x^{2\cdot 2^k}+x^{3\cdot 2^k}+\ldots)$$

$$= \sum_{s=0}^{\infty} G(s)x^s,$$

where $G(0)=1$ and $G(s)$ (for s greater than 0) is the number of ways in which the positive integer s may be separated into like or distinct summands each of which is a power of 2.

We have readily

$$(1-x)\sum_{s=0}^{\infty} G(s)x^s = (1-x)P(x) = P(x^2) = \sum_{s=0}^{\infty} G(s)x^{2s};$$

whence

$$G(2s+1) = G(2s) = G(2s-1)+G(s), \qquad (A)$$

as one readily verifies by equating coefficients of like powers of x. From this we have in particular

$$G(0)=1, \quad G(1)=1, \quad G(2)=2, \quad G(3)=2,$$

$$G(4)=4, \quad G(5)=4, \quad G(6)=6, \quad G(7)=6.$$

Thus in (A) we have recurrence relations by means of which we may readily reckon out the values of the number theoretic function $G(s)$. Thus we may determine the number of ways in which a given positive integer s may be represented as a sum of powers of 2.

We have given this example as an elementary illustration of the analytical theory of numbers, that is, of that part of the theory of numbers in which one employs (as above) the theory of a continuous variable or some analogous theory in order to derive properties of sets of integers. This general subject has been developed in several directions. For a systematic account of it the reader is referred to Bachmann's Analytische Zahlentheorie.

§ 45. Diophantine Equations

If $f(x, y, z, \ldots)$ is a polynomial in the variables $x, y, z, \ldots$ with integral coefficients, then the equation

$$f(x, y, z, \ldots)=0$$

is called a Diophantine equation when we look at it from the point of view of determining the integers (or the positive integers) $x, y, z, \ldots$ which satisfy it. Similarly, if we have several such functions $f_i(x, y, z, \ldots)$, in number less than the number of variables $x, y, z, \ldots$, then the set of equations

$$f_i(x, y, z, \ldots)=0, \quad i=1, 2, \ldots,$$

is said to be a Diophantine system of equations. Any set of integers $x, y, z, \ldots$ which satisfies the equation [system] is said to be a solution of the equation [system].

We may likewise define Diophantine inequalities by replacing the sign of equality above by the sign of inequality. But little has been done toward developing a theory of Diophantine inequalities. Even for Diophantine equations the theory is in a rather fragmentary state.

In the next two sections we shall illustrate the nature of the ideas and the methods of the theory of Dipohantine equations by developing some of the results for two important special cases.

§ 46. Pythagorean Triangles

Definitions. If three positive integers x, y, z satisfy the relation

$$x^2+y^2=z^2 \tag{1}$$

they are said to form a Pythagorean triangle or a numerical right triangle; z is called the hypotenuse of the triangle and x and y are called its legs. The area of the triangle is said to be $\frac{1}{2}xy$.

We shall determine the general form of the integers x, y, z, such that equation (1) may be satisfied. Let us denote by ν the greatest common divisor of x and y in a particular solution of (1). Then ν is a divisor of z and we may write

$$x=\nu u, \quad y=\nu v, \quad z=\nu w.$$

Substituting these values in (1) and reducing we have

$$u^2+v^2=w^2, \tag{2}$$

where u, v, w are obviously prime each to each, since u and v have the greatest common divisor 1.

Now an odd square is of the form $4k+1$. Hence the sum of two odd squares is divisible by 2 but not by 4; and therefore the sum of two odd squares cannot be a square. Hence one of the numbers u, v is even. Suppose that u is even and write equation (2) in the form

$$u^2=(w-v)(w+v). \tag{3}$$

Every common divisor of $w-v$ and $w+v$ is a divisor of their difference $2v$. Therefore, since w and v are relatively prime, it follows that 2 is the greatest common divisor of $w-v$ and $w+v$. Then from (3) we see that each of these numbers is twice a square, so that we may write

$$w-v=2b^2, \quad w+v=2a^2$$

where a and b are relatively prime integers. From these two equations and equation (3) we have

$$w=a^2+b^2, \quad v=a^2-b^2, \quad u=2ab. \tag{4}$$

Since u and v are relatively prime it is evident that one of the numbers a, b is even and the other odd.

The forms of u, v, w given in (4) are necessary in order that (2) may be satisfied. A direct substitution in (2) shows that this equation is indeed satisfied by these values. Hence we have in (4) the general solution of (2) where u is restricted to be even. A similar solution would be obtained if v were restricted to be even. Therefore *the general solution of* (1) *is*

$$x=2\nu ab, \quad y=\nu(a^2-b^2), \quad z=\nu(a^2+b^2)$$

and

$$x=\nu(a^2-b^2), \quad y=2\nu ab, \quad z=\nu(a^2+b^2)$$

where a, b, ν are arbitrary integers except that a and b are relatively prime and one of them is even and the other odd.

By means of this general solution of (1) we shall now prove the following theorem:

I. *There do not exist integers m, n, p, q, all different from zero, such that*

$$q^2+n^2=m^2, \quad m^2+n^2=p^2. \tag{5}$$

It is obvious that an equivalent theorem is the following:

II. *There do not exist integers m, n, p, q, all different from zero such that*

$$p^2+q^2=2m^2, \quad p^2-q^2=2n^2. \tag{6}$$

Obviously, we may without loss of generality take m, n, p, q to be positive; and this we do.

The method of proof is to assume the existence of integers satisfying equations (5) and (6) and to show that we are thus led to a contradiction. The argument we give is an illustration of Fermat's famous method of "infinite descent."

If any two of the numbers p, q, m, n have a common prime factor t, it follows at once from (5) and (6) that all four of them have this factor. For, consider an equation in (5) or in (6) in which these two numbers occur; this equation contains a third number, and it is readily seen that this third number is divisible by t. Then from one of the equations containing the fourth number it follows that this fourth number is divisible by t. Now let us divide each equation of system (6) through by t^2; the resulting system is of the same form as (6). If any two numbers in this resulting system have a common prime factor t_1, we may divide through by t_1^2; and so on. Hence if a pair of simultaneous equations (6) exists then there exists a pair of equations of the same form in which no two of the numbers m, n, p, q have a common factor other than unity. Let this system of equations be

$$p_1^2+q_1^2=2m_1^2, \quad p_1^2-q_1^2=2n_1^2. \tag{7}$$

From the first equation in (7) it follows that p_1 and q_1 are both even or both odd; and, since they are relatively prime, it follows that they are both odd. Evidently $p_1>q_1$. Then we may write

$$p_1=q_1+2\alpha,$$

where α is a positive integer. If we substitute this value of p_1 in the first equation of (7), the result may readily be put in the form

$$(q_1+\alpha)^2+\alpha^2=m_1^2. \tag{8}$$

Since q_1 and m_1 have no common prime factor it is easy to see from this equation that α is prime to both q_1 and m_1, and hence that no two of the numbers $q_1+\alpha$, α, m_1 have a common factor.

Now we have seen that if a, b, c are positive integers no two of which have a common prime factor, while

$$a^2+b^2=c^2,$$

then there exist relatively prime integers r and s, $r>s$, such that

$$c=r^2+s^2, \quad a=2rs, \quad b=r^2-s^2$$

or

$$c=r^2+s^2, \quad a=r^2-s^2, \quad b=2rs.$$

Hence from (8) we see that we may write

$$q_1+\alpha=2rs, \quad \alpha=r^2-s^2 \tag{9}$$

or

$$q_1+\alpha=r^2-s^2, \quad \alpha=2rs. \tag{10}$$

In either case we have

$$p_1^2-q_1^2=(p_1-q_1)(p_1+q_1)=2\alpha\cdot 2(q_1+\alpha)=8rs(r^2-s^2).$$

If we substitute in the second equation of (7) and divide by 2 we have

$$4rs(r^2-s^2)=n_1^2.$$

From this equation and the fact that r and s are relatively prime it follows at once that r, s, r^2-s^2 are all square numbers; say,

$$r=u^2, \quad s=v^2, \quad r^2-s^2=w^2.$$

Now $r-s$ and $r+s$ can have no common factor other than 1 or 2; hence from

$$w^2=(r^2-s^2)=(r-s)(r+s)=(u^2-v^2)(u^2+v^2)$$

we see that either

$$u^2+v^2=2w_1^2, \quad u^2-v^2=2w_2^2 \tag{11}$$

or

$$u^2+v^2=w_1^2, \quad u^2-v^2=w_2^2.$$

And if it is the latter case which arises, then

$$w_1^2+w_2^2=2u^2, \quad w_1^2-w_2^2=2v^2. \tag{12}$$

Hence, assuming equations of the form (6) we are led either to equations (11) or to equations (12); that is, we are led to new equations of the form with which we started. Let us write the equations thus:

$$p_2^2+q_2^2=2m_2^2, \quad p_2^2-q_2^2=2n_2^2; \tag{13}$$

that is, system (13) is identical with that one of systems (11), (12) which actually arises.

Now from (9) and (10) and the relations $p_1=q_1+2\alpha$, $r>s$, we see that

$$p_1=2rs+r^2-s^2>2s^2+r^2-s^2=r^2+s^2=u^4+v^4.$$

Hence $u<p_1$. Also,

$$w_1^2 \leqq w^2 \leqq r+s<r^2+s^2.$$

Hence $w_1<p_1$. Since u and w_1 are both less than p_1 it follows that p_2 is less than p_1. Hence, obviously, $p_2<p$. Moreover, it is clear that all the numbers p_2, q_2, m_2, n_2 are different from zero.

From these results we have the following conclusion: If we assume a system of the form (6) we are led to a new system (13) of the same form; and in the new system p_2 is less than p.

Now if we start with (13) and carry out a similar argument we shall be led to a new system

$$p_3^2+q_3^2=2m_3^2, \quad p_3^2-q_3^2=2n_3^2,$$

with the relation $p_3<p_2$; starting from this last system we shall be led to a new one of the same form, with a similar relation of inequality; and so on *ad infinitum*. But, since there is only a finite number of positive integers less than the given positive integer p this is impossible. We are thus led to a contradiction; whence we conclude at once to the truth of II and likewise of I.

By means of theorems I and II we may readily prove the following theorem:

III. *The area of a numerical right triangle is never a square number.*

Let the sides and hypotenuse of a numerical right triangle be u, v, w, respectively. The area of this triangle is $\frac{1}{2}uv$. If we assume this to be a square number t^2 we shall have the following simultaneous Diophantine equations

$$u^2+v^2=w^2, \quad uv=2t^2. \tag{14}$$

We shall prove our theorem by showing that the assumption of such a system leads to a contradiction.

If any two of the numbers u, v, w have a common prime factor p then the remaining one also has this factor, as one sees readily from the first equation in (14). From the second equation in (14) it follows that t also has the same factor. Then if we put $u=pu_1$, $v=pv_1$, $w=pw_1$, $t=pt_1$, we have

$$u_1^2+v_1^2=w_1^2, \quad u_1v_1=2t_1^2,$$

a system of the same form as (14). It is clear that we may start with this new system and proceed in the same manner as before, and so on, until we arrive at a system

$$\bar{u}^2+\bar{v}^2=\bar{w}^2, \quad \bar{u}\,\bar{v}=2\bar{t}^2, \qquad (15)$$

where $\bar{u}$, $\bar{v}$, $\bar{w}$ are prime each to each.

Now the general solution of the first equation (15) may be written in one of the forms

$$\bar{u}=2ab, \quad \bar{v}=a^2-b^2, \quad \bar{w}=a^2+b^2,$$
$$\bar{u}=a^2b^2, \quad \bar{v}=2ab, \quad \bar{w}=a^2+b^2.$$

Then from the second equation in (15) we have

$$\bar{t}^2=ab(a^2-b^2)=ab(a-b)(a+b).$$

It is easy to see that no two of the numbers a, b, $a-b$, $a+b$ in the last member of this equation have a common factor; for, if so, $\bar{u}$ and $\bar{v}$ would have a common factor, contrary to hypothesis. Hence each of these four numbers is a square. That is, we have equations of the form

$$a=m^2, \quad b=n^2, \quad a+b=p^2, \quad a-b=q^2;$$

whence $$m^2-n^2=q^2, \quad m^2+n^2=p^2.$$

But, according to theorem I, no such system of equations can exist. That is, the assumption of equations (14) leads to a contradiction. Hence the theorem follows as stated above.

§ 47. The Equation $x^n+y^n=z^n$.

The following theorem, which is commonly known as Fermat's Last Theorem, was stated without proof by Fermat in the seventeenth century:

If n is an integer greater than 2 there do not exist integers x, y, z, all different from zero, such that

$$x^n+y^n=z^n. \qquad (1)$$

No general proof of this theorem has yet been given. For various special values of n the proof has been found; in particular, for every value of n not greater than 100.

In the study of equation (1) it is convenient to make some preliminary reductions. If there exists any particular solution of (1) there exists also a solution in which x, y, z are prime each to each, as one may show readily by the method employed in the first part of § 46. Hence in proving the impossibility of equation (1) it is sufficient to treat only the case in which x, y, z are prime each to each.

Again, since n is greater than 2 it must contain the factor 4 or an odd prime factor p. If n contains the factor p we write $n=mp$, whence we have

$$(x^m)^p+(y^m)^p=(z^m)^p.$$

If n contains the factor 4 we write $n=4m$, whence we have

$$(x^m)^4+(y^m)^4=(z^m)^4.$$

From this we see that in order to prove the impossibility of (1) in general it is sufficient to prove it for the special cases when n is 4 and when n is an odd prime p. For the latter case the proof has not been found. For the former case we give a proof below. The theorem may be stated as follows:

I. *There are no integers x, y, z, all different from zero, such that*

$$x^4+y^4=z^4.$$

This is obviously a special case of the more general theorem:

II. *There are no integers p, q, α, all different from zero, such that*

$$p^4 - q^4 = \alpha^2. \qquad (2)$$

The latter theorem is readily proved by means of theorem III of § 46. For, if we assume an equation of the form (2), we have

$$(p^4 - q^4)p^2q^2 = p^2q^2\alpha^2. \qquad (3)$$

But, obviously,

$$(2p^2q^2)^2 + (p^4 - q^4)^2 = (p^4 + q^4)^2. \qquad (4)$$

Now, from (3) we see that the numerical right triangle determined by (4) has its area $p^2q^2(p^4 - q^4)$ equal to the square number $p^2q^2\alpha^2$. But this is impossible. Hence no equation of the form (2) exists.

EXERCISES

1. Show that the equation $\alpha^4 + 4\beta^4 = \gamma^2$ is impossible in integers α, β, γ, all of which are different from zero.

2. Show that the system $p^2 - q^2 = km^2$, $p^2 + q^2 = kn^2$ is impossible in integers p, q, k, m, n, all of which are different from zero.

3*. Show that neither of the equations $m^4 - 4n^4 = \pm t^2$ is possible in integers m, n, t, all of which are different from zero.

4*. Prove that the area of a numerical right triangle is not twice a square number.

5*. Prove that the equation $m^4 + n^4 = \alpha^2$ is not possible in integers m, n, α all of which are different from zero.

6*. In the numerical right triangle $a^2 + b^2 = c^2$, not more than one of the numbers a, b, c is a square.

7. Prove that the equation $x^{2k} + y^{2k} = z^{2k}$ implies an equation of the form $m^k + n^k = 2^{k-2}t^k$.

8. Find the general solution in integers of the equation $x^2 + 2y^2 = t^2$.

9. Find the general solution in integers of the equation $x^2 + y^2 = z^4$.

10. Obtain solutions of each of the following Diophantine equations:

$$x^3 + y^3 + z^3 = 2t^3,$$
$$x^3 + 2y^3 + 3z^3 = t^3,$$
$$x^4 + y^4 + 4z^4 = t^4,$$
$$x^4 + y^4 + z^4 = 2t^4.$$

INDEX